INSCRUTABLE
EARTH

also by Ron Parker

THE SHEEP BOOK

INSCRUTABLE EARTH

Explorations into the Science of Earth

RONALD B. PARKER

CHARLES SCRIBNER'S SONS · NEW YORK

Copyright © 1984 Ronald B. Parker

Library of Congress Cataloging in Publication Data

Parker, Ronald B.
  Inscrutable earth.

  Bibliography: p.
  Includes index.
  1. Geology.   I. Title.
QE31.P37 1984   550   84-10681
ISBN 0-684-18173-8

1 3 5 7 9 11 13 15 17 19   V/C   20 18 16 14 12 10 8 6 4 2

Printed in the United States of America.

Contents

*Dedicated to my parents who created me,
and to the Creator who gave us this earth and
its creatures:
the ultimately interesting puzzle*

Acknowledgments

The language and the concepts used in this book have been mulled over by a number of readers. Their help has been invaluable in making the final product more readable as well as scientifically more sound. I am especially grateful for the numerous warm get-togethers with Heinrich Toots of C. W. Post College's Department of Geology and Geography and with Mike Voorhies of the University of Nebraska State Museum, both of whom gave unsparingly of their time and genius. My wife Teresa and our friend Kathy Barber both acted as able amateurs, playing the role of intelligent nonspecialists who pointed out many instances where I assumed that the reader knew a bit more than could be reasonably expected. I, of course, take full responsibility for any unaccustomed ideas presented herein that may surprise some traditionalists in the area of earth sciences.

And some rin up hill and down dale,
knapping the chucky stanes to pieces wi' hammers,
like sae many road makers run daft.
They say it is to see how the warld was made

Sir Walter Scott, St. Roman's Well, *1824*

Introduction

G EOLOGY IS THE ROOT OF THE FAMILY TREE OF SCIENCES, ALL of which have branched out from that sturdy beginning. As such, the study of the earth has been the starting point for disciplines that are now considered by their practitioners to be more basic. Yet chemistry started as a branch of mineralogy, and physics is nothing more than the study, through such specialties as mechanics, thermodynamics, and microphysics, of a narrow range of properties of earth materials. Just as today's astrophysicists ponder the origins of the cosmos, so the cosmogenists of the turn of the seventeenth century pondered the origin of the earth. Biology and paleontology were almost indistinguishably intermixed in their early days, and meteorology and oceanography are children of the science of the earth as a whole. Astronomy grew out of the mythology of astrology and became associated with geology early on. Today geologists utilize the results of research in these various derived disciplines to

1

advance our understanding of this most inaccessible part of the universe that we dwell upon, a place we know only from its surface features, a few shallow holes, and the rumblings and heat from its interior. Earth is a place that in many ways is less well known than distant stars, for stars can be studied by techniques that cannot be used to study our home. For example, the chemical composition of stars can be determined from the light and other radiation that they emit; their temperatures measured by a related method; and their magnetic field, density, and internal movement assessed. We cannot measure Earth's properties so easily. Our earth is difficult to understand. Indeed, there are properties of our planet that we will never measure, past events that we will never be able to reconstruct, and grand changes that our short-term observations will not detect. The earth is unknowable, inscrutable.

Geology is commonly misunderstood by lay persons. The general public imagines that most geologists work for oil companies, trying to locate new petroleum deposits. People who have taken a beginning geology course in college or perhaps an earth science course in secondary school tend to think that geologists identify rocks and minerals and that paleontologists identify fossils. I can't blame them for their misconceptions because the teachers of such courses—the kind called "rocks for jocks" in many colleges and universities—commonly inform the class when it first meets that geology is a descriptive science, nothing more than a catalog of observations. The students find little to dissuade them from that view during the term, for they soon discover that the text is a body of descriptions and definitions of terms that is systematically "mined" for examination questions. The lectures rarely reveal many of the great theories of earth science, nor do they allude to the vast, unsolved problems of the field.

Fortunately, there are some colleges and universities where this is not the case. At some enlightened institutions—and their enlightenment bears no simple relationship to size, antiquity, or reputation—the beginning geology courses are taught by the top professors instead of being delegated to the staff members with the least clout—the newest or the oldest and stalest. Students who are fortunate enough to attend one of the few excellent elementary geology courses soon discover that geology is no more a descriptive science than any other field. Geology is a study of the nature of the earth in *all* its ramifications.

Geology is a field of scholarship that attempts to probe into the past, a past that is billions of years old. The earth is like a huge history book with pages that are missing, illegible, written in unknown languages, filled with surprises and puzzles. The pages that we can read and understand are so fascinating that the book demands to be read and reread in new ways so that more might be discovered. Geology involves the piecing together of bits of evidence to test a theory. Every theory leads to the search for new basic data and in many cases to new and different hypotheses. Geology is the study of the building of continents and mountain ranges, the evolution of life, the rumbling of earthquakes, and the bursting forth of volcanoes. Geology is also the study of the atomic arrangement of crystals, the nature of billion-year-old microfossils, the shape of sand grains, and the composition of a fossil's tooth.

Geology can also be fun. Many scientists, like professionals in other fields, act as if their discipline can be seriously entered into only by those who act as if it were all drudgery. Yet during over three decades of close association with scientists, it has been my experience that the top people in a given field are the ones most likely to be talking and laughing together about their work at a professional gathering.

I have tried to pass on some of the excitement of earth science in the essays that follow. I also hope to give you a sense of the scope of geology and its sister science, paleontology. This is not a textbook in the usual sense, for I have tried to avoid using a technical term where a common English word will do. I have not, however, scrimped on content: Some of the topics I have covered have previously been treated only in fairly obscure publications. I have also introduced quite a few original concepts and have tied together some seemingly unrelated ideas to suggest solutions to geological problems. Because much of this material has not been published previously, even some readers who are knowledgeable about earth sciences will find some fresh ideas to chew upon. The first few essays lay the groundwork of geology—the questions of the chemical composition, age, and energy structure of the earth—and are followed by an examination of the nature and origin of the oceans and the atmosphere. The remaining essays, which cover a variety of different topics, large and small, are a tempting look at the periphery of earth science research. In the last essay I have tried to tie together two concepts—that of an evolving earth and that of an earth that is characterized by cycles—into a coherent whole, for in our fascination with cycles we must not forget that the earth changes unrecoverably with the unidirectional passage of time. That continuous change is the cloth on which the cycles are mere embroidery. To reveal the pattern we must unravel the tapestry, and we must do so if we are to comprehend the processes that made our earth—our inscrutable earth.

Chapter 1

Assaying the Earth

*I*N THE SIXTH CENTURY B.C., THE GREEK PHILOSPHER THALES PRO-
posed that all substances derived from water, whereas a
fellow philosopher, Anaximenes, thought that air was the
essential element. Later, in the fifth century B.C., yet another
philosopher, Heraclitus, insisted that fire was the important
factor. Still later, in the fourth century B.C., Aristotle straight-
ened everyone out with what seemed to be the hardly less
ludicrous theory that all substances in the world were composed of various proportions of earth, air, water, and fire.
Before we are too critical of our predecessors, we should recognize that their theories were not really all that silly; they
were just not detailed enough to be of much practical application. If we take earth, air, and water to mean solids, gases,
and liquids, then any object can be said to be composed of a

For more information, see references 1, 16, 17, 29, 48, 66, 68, 75, 89, and
90 in *Further Reading*.

combination of those materials. Plants, for example, are mostly water plus carbon dioxide mixed with minor amounts of solid mineral matter—in other words water, air, and a trace of earth. Rocks are solid mineral matter also containing minor amounts of gases and water that can be driven off by heating. A glass of sparkling mineral water is water plus air. And so it goes.

It was reasonable for the ancients to regard fire as a primary element since they had observed that heat was required to form some things and was given off from others. For example, they found that if quartz sand, crushed feldspar, and soda were mixed and heated to a sufficiently high temperature, they melted to form glass. Therefore glass could be said to be composed of earth and fire. Modern-day ceramists would confirm this observation although they would say that the glass was at a higher energy state than the raw components, the energy (derived from heat) being the "fire" of the old days. A piece of firewood, when burned, gives off heat and gases (including water vapor), and some ash remains. Therefore wood could be said to be made of fire, air, water, and earth. Ridiculous? Not when you consider that the burnable carbohydrates and other carbon compounds in wood are formed from water and carbon dioxide with the help of energy from the sun that drives the process of photosynthesis. The solar energy (fire) is stored in the chemical bonds of the plant tissues.

The only real problem with the old concepts was that they lacked sufficient resolution. What might have been adequate for ivory tower philosophers was of little practical worth to those of an applied science persuasion. The early equivalents of today's chemical engineers knew that the ash produced by heating seashells was different from that produced by burning wood or by burning salt-tolerant plants, and they manufactured lime, potash, and soda ash, respectively, from those raw materials. Those early manufacturers therefore knew that there had to be different kinds of earth, air, and water.

Long before Roman times—as early as the Bronze Age, around 9,000–8,000 B.C.—miners had learned how to extract useful materials, especially metals, from rocks and in so doing developed simple methods of chemical analysis to evaluate the potential of ores. In his treatise on the Red Sea, *Periplus Rubri Maris*, the Greek historian and geographer Agatharchides of Samos (181–146 B.C.), described a process for purification of gold that was used at a gold deposit on the Nile. The impure element was placed together with lead, salt, a little tin, and some barley or bran into a crucible which was covered and then heated for several days. After cooling, a button of nearly pure gold was removed from the crucible. This method is nearly identical to the fire assay used for purifying gold as late as the first part of this century. Thus the quantitative analysis of rocks was born long ago.

From about the fifth century on, alchemists, engaged in a fruitless search for the philosopher's stone, that imaginary substance that had the power to transmute base metals into gold. As we know, the mystical practice of alchemy never did find the way to make gold from less noble materials (although alchemists should have had the sense to realize that if the process had been discovered, the value of gold would have plummeted, so the search was doomed from the start) but in their search for magical things the alchemists discovered a lot of basic chemistry. The list of four fundamental components was thus expanded to include things like ochres, salts, clays, alums, ashes, and acids.

From ancient times to the early part of the twentieth century, chemists have been largely concerned with breaking things down into their component parts, that is, chemical elements. Rocks and minerals were the favorite raw materials of the chemists, and by the middle of the nineteenth century, the Swedish chemist Berzelius and his colleagues had accumulated a vast amount of data on the composition of minerals. Fifty years later, enough was known about the elements to allow the Russian chemist Mendeléev to construct his periodic table, a classification of the elements by their chemical properties. Mendeléev's periodic table contained gaps where according to his theory yet undiscovered elements should fit, and chemists scurried to their mineral collections to seek out the missing rascals. In 1923 a new technique—X rays—was used by a pair of physicists in Copenhagen to discover the element hafnium (named after the city's old name, Hafnia) in the mineral zircon. Hafnium was indeed located where Mendeléev's theory predicted it would be, for in his periodic table was a gap for an element that would be chemically similar to zirconium, a major element in zircon. The days of what might be called the exploration phase of chemistry were for all intents finished, but in this phase analytical techniques were developed to a routine rather than an art.

The nineteenth century was a time of great discovery in geology. Armed with analytical chemistry and microscopes that used polarized light (which made it possible to measure many properties of mineral crystals that cannot be studied by ordinary light), geologists set forth to describe rocks in a thoroughly modern way. Perhaps inspired by the then-current fashion of breaking things down to as many subdivisions as possible, many workers succumbed to the temptation to assign a unique name to every slightly different rock they encountered in their studies. It was a time of splitters rather than lumpers, and the literature of petrology (the study of rocks) became burdened with many names that would prove useless in the years ahead. It was a time of collecting and describing data with little thought as to why.

The grand champion splitter was Professor Waldemar Brögger of Norway who, around the turn of the present century, assigned new names to virtually every rock that crossed his desk. It came as a real shock to me to visit some of Brögger's type localities (a type locality is a place where a described rock type is first collected) and to find that more than one of his rock names referred to a single outcropping poking through the soil of the farm that gave it its name (petrographic types are usually named for places). It was sobering to find that some of the famous names of petrography, as the practice of describing rocks is called, came from outcrops of less than a square foot in area.

Because of Brögger and other like him, the literature of

geology was filled with descriptions of individual rocks, but at the beginning of the present century there was really very little good quantitative information about the average composition of large areas of the earth's crust, the outermost layer of the solid earth (about 50 kilometers, or 30 miles, thick under the continents). Geologists and an emerging breed that called themselves geochemists realized that what was needed was analyses of common rocks, not rare and unusual ones. What was needed was some idea of the composition of the earth as a whole.

Estimating the composition of the earth's crust is not a simple matter, especially when compared with estimating the composition of the atmosphere or the hydrosphere (the waters of the earth). In the case of the atmosphere, the analysis of almost any air sample gives a fair estimate of the composition of the entire gaseous envelope because the constant state of mixing that we call weather keeps it relatively homogeneous. Over 99 percent of the atmosphere is a mixture of nitrogen, oxygen, and argon. Carbon dioxide is a very minor component, and neon and krypton, are the only other gases present in amounts exceeding a part per million (ppm).

Characterizing the hydrosphere is also relatively easy. About 95 percent of the earth's waters are in the oceans, and their compositions, while not as uniform as those of the atmosphere, do not range widely. In the open ocean the percentage of dissoved solids, or salinity, is between 3 and 4 percent, with most values between 3.4 and 3.7. There are waters that are much more or much less saline, but they are of little quantitative importance.

In contrast, the crust of the solid earth is highly inhomogeneous; it does not mix except on a scale of billions of years. The problem of determining the composition of this crust is one of sampling, as the statisticians would call it. It is the same problem faced by a pollster who wants to get a reasonable estimate of the behavior of a group of voters or consumers by questioning a small number of the larger population or by the quality control person who wants to ensure that most of the nuts and bolts or automobiles that come out of the factory are satisfactory without having to check every single one.

A geochemist designing a project to determine the composition of the earth's crust from scratch might collect samples randomly in three dimensions and analyze them all. Yet the third dimension—depth—is difficult to sample, partly because drill holes are expensive and partly because even the deepest ones (about 7 kilometers, or 4 miles) only penetrate about 10 percent of the crust's total thickness. Alternately, the geochemist might collect samples in two dimensions on a regular grid pattern to get a good estimate of the upper crust. This has been done for the granitic rocks of southern California by Alex Baird and his co-workers at Pomona College. As a practical matter, however, a worldwide project of this sort is not feasible even today, much less at the start of the twentieth century, both because of the cost and because bedrock, the name given to solid, crustal rock, is not everywhere exposed for sampling because of cover by water, soil, weathered residual rock, and the like.

To estimate crustal composition, scientists had to select the data they already had in such a way that they felt assured, a priori, that their sample would be representative of the whole. In other words, they relied on their hunches. The first earth scientists to do this were F. W. Clarke and H. S. Washington of the U.S. Geological Survey. They had thousands of analyses in their files to use as their data base. They made the geological assumption that all rocks in the crust were either igneous rocks or were derived from igneous rocks, so that it was not necessary to include analyses of the quantitatively less important sedimentary or metamorphic types.*

For their study, published in 1924, they selected over five thousand analyses of igneous rocks that Washington (the

---

*Igneous rocks (such as basalt, gabbro, and many granites) are those formed by the solidifying of molten liquids. Sedimentary rocks (such as sandstone, shale, and limestone) are those formed by an accumulation of particles or dissolved solids. Metamorphic rocks (such as schist, gneiss, and marble) are those formed by the change of some previous rock type, usually in response to new conditions of temperature and/or pressure.

geologist of the two, Clarke being a chemist) considered reliable and averaged them. They also made the assumption that the unusual compositions of bizarre rock types represented in their files would effectively balance one another and that averaging all values would yield a valid crustal composition estimate. They tested their results by breaking them down to averages for continents or similarly sized large units and found the averages to be in general agreement, so they concluded that their assumptions had been valid.

Though they found that oxygen (chemical symbol O) was the most abundant element in the crust, they did not list oxygen separately but, following the custom of their day, gave the results in terms of oxides of the elements. They found that the crust contained, in percentages according to weight, roughly 60 percent silica (SiO_2), 15 percent alumina (Al_2O_3), 7 percent iron oxides (Fe_2O_3 + FeO), and 3.5 percent magnesia (MgO), with another half dozen oxides in appreciable amounts (see the appendix for chemical symbols if you are unfamiliar with them). Clarke and Washington used only analyses of samples collected on the continents, so their estimate was actually an estimate of the composition of the continental crust rather than the entire crust, which would have required samples from oceanic basins.

Clarke and Washington's estimates had a mixed reception. Most critics objected to their assumptions, particularly the assumption that averaging analyses of a mixed bag of rock types would yield an average crustal rock composition. One critic was Victor Moritz Goldschmidt, a Swiss who had moved to Norway as a lad, had later become a student of Brögger, and who in 1914 was appointed professor and head of the Mineralogical Institute at Oslo. Goldschmidt, sometimes called the father of modern geochemistry, was steeped in the igneous tradition that dominated European thinking of the time. According to this tradition, igneous rocks were the really fundamental rocks of the crust whereas metamorphic and sedimentary rocks were of little importance. Still, Goldschmidt did not agree with using analyses of unusual igneous rock types for the task of estimating crustal composition. Therefore he made no use of analyses collected by Brögger but proposed that one could get a good average sample of the exposed rocks of Norway by analyzing samples of glacial debris, material that had been already collected by a long-gone ice sheet. The distinguished Goldschmidt (who had the misfortune of strongly resembling Oliver Hardy of movie fame) was not a field geologist at heart and did not go outdoors and collect samples of glacial material for analysis, though he is credited with doing so in a number of textbooks. Instead he went to his bookshelf and extracted some soil analyses from a government publication and used these for his averaging, reasoning that the soils were so young that their original composition would not have changed a great deal. His assumption was justified to the extent that his average of seventy-seven samples agreed reasonably well with Clarke and Washington's except that his average was lower in calcium and very much lower in sodium. This was to be expected; calcium would have been utilized by plants in tilled soils, and sodium is relatively soluble and thus susceptible to leaching. I do find it surprising that Goldschmidt didn't just send an assistant to collect some better samples for analysis in the excellent laboratories at Oslo, but he must have had his reasons. He did, however, use a different sampling technique that was sounder than Clarke and Washington's.

Across the Baltic in Finland, the great geologist J. J. Sederholm was not happy with Clarke and Washington's use of unusual rock types either. He therefore computed his own average of the Finnish crust by using a geologic map of his homeland and weighting analyses from the files of the Geological Survey of Finland of the various major rock types according to the area of their exposure. His 1925 average showed more silica (67%) and less iron and magnesium than other estimates, reflecting the fact that Finnish crust is considerably more granitic than the continental crust as a whole.

In 1955 Arie Poldervaart of Columbia University applied Sederholm's method on a worldwide scale in three dimensions. He recognized that the oceanic crust was different from the continental crust, took into account the thin sedimentary veneer that covered major parts of the crust, and included mountain belts as a subdivision. Poldervaart's average contains less silica (55%) and more calcium than the previous estimates. He also demonstrated that sedimentary rocks have a composition very different from that of the average igneous rocks from which, according to Clarke and Washington, they were supposed to be derived. It is doubtful that Poldervaart's values will be substantially altered in the future. In any case, if you look at many current texts and other publications, you'll find that authors typically recycle Clarke and Washington's old data instead of using more modern values—whether through ignorance or laziness is open to debate.

Lest you get the impression that the question of crustal composition has been solved, let me remind you of the scale of the problem. As I mentioned earlier, the deepest boreholes from which we can obtain samples are only about 7 kilometers (4 miles) deep, and the continental crust is about 50 kilometers (30 miles) thick. Moreover, the crust accounts for only about 1 percent of the total mass of the earth. Underneath it, to a depth of about 3,000 kilometers (1,900 miles), is a shell called the mantle, which in turn surrounds a central core with a radius of about 3,500 kilometers (2,200 miles). Put another way, the radius of the entire earth is about a thousand times greater than our deepest borehole. Estimating the composition of the whole earth is *really* a problem.

We have been able to characterize mantle rocks largely from the way in which they transmit earthquake vibrations. We do get a few samples of the mantle at places where it has been shoved over the oceanic crust or where samples have been delivered to the surface by cooperative volcanoes. From these paltry fragments we know that the mantle is richer in iron and magnesium and poorer in silica and alkalies (such as sodium and potassium) than the crust.

The earth's core is presumed to consist largely of metallic iron and nickel and to have an outer molten part and an inner solid part. This presumption is based on earthquake studies and on the fact that the total density of the earth can be explained only if we assume that the core is very dense, and those elements fit the density requirement very nicely.

It you sent off a piece of rock to an analytical laboratory, in due time you would receive a bill for one hundred dollars or more and, in keeping with the old-fashioned tradition of listing elements as if they were combined with oxygen, a list of the amounts of the fourteen oxides present in the rock. A curious amateur would notice that iron is listed twice, once as FeO and again as Fe_2O_3. If we rewrite FeO as Fe_2O_2 it may make it clear that there are two oxygen atoms for every two iron atoms in FeO whereas there are three for every two in Fe_2O_3. In other words, rocks containing both FeO and Fe_2O_3 could have held a bit more oxygen—if that FeO could have been fully oxidated to Fe_2O_3. Practically all rocks in the crust have the potential for holding a bit more oxygen than they actually do, a characteristic that becomes more pronounced with depth.

Water is also listed twice in the analysis to distinguish between the water that is just moisture and that which is chemically combined in crystals. The latter is shown as H_2O + 105 degrees which means that it is driven off, or vaporized, above the boiling point of water, 100°C. Mineralogists call this water hydroxyl, and we'll have reason to look at it again in Chapter 7.

It is far more informative to list analyses in terms of elements, as in the table below, which is an average of crustal compositions given by Brian Mason of the U.S. National Museum and A. P. Vinogradov of the Vernadski Institute of the Academy of Sciences of the USSR. A selection of minor or trace elements is also given, and for this reason the table is presented in terms of parts per million (ppm) instead of percent. To convert to percent just move the decimal point four places to the left. Those of you who are not used to thinking in ppm can get an intuitive feeling for it by recognizing that it is equivalent to grams per metric ton. A small import car weighs about a metric ton (roughly 2,200 pounds) and a gram is approximately the weight of three ordinary aspirin tablets.

AVERAGE COMPOSITION OF THE CRUST, PPM, INCLUDING SELECTED MINOR ELEMENTS

element		element	
oxygen (O)	468,000	boron (B)	11
silicon (Si)	286,100	carbon (C)	215
aluminum (Al)	80,900	nitrogen (N)	20
iron (Fe)	48,250	fluorine (F)	642
calcium (Ca)	32,950	sulfur (S)	365
sodium (Na)	26,650	chlorine (Cl)	150
potassium (K)	25,450	chromium (Cr)	92
magnesium (Mg)	19,800	nickel (Ni)	66
titanium (Ti)	4,450	copper (Cu)	51
hydrogen (H)	1,400*	zinc (Zn)	76
phosphorus (P)	990	strontium (Sr)	358
manganese (Mn)	975	lead (Pb)	14

*Mason's value only; Vinogradov doesn't estimate H.

By weight, the first eight elements account for 98.8 percent of the crust, with no other elements present in an amount even as great as 1 percent and most of them in far smaller amounts. An element such as boron may seem pretty minor at 11 ppm (0.0011%), but even boron is practically abundant when compared with mercury (0.08 ppm) or platinum (less than 0.01 ppm or 0.000001%). Small wonder that platinum commands a fancy price even before it is made into jewelry and marked up still further. For a more complete table of the abundance of the elements in the crust, see the appendix.

It is surprising to lay persons, and even to most geologists, that oxygen makes up almost half the weight of the crust. If we were to look at the oxygen content in terms not of weight but of the number of atoms, we would find that it accounts for 63 percent of the crust. Oxygen is relatively large when compared with other atoms, especially the abundant ones like silicon and aluminum, and if we convert the percentages to volume percent we find that oxygen comprises an astonishing 94 percent of the crust. If this doesn't surprise you, then you have doubtless forgotten that oxygen makes up only about 21 percent of the atmosphere, meaning that the solid crust contains about four and a half times as much oxygen as the air we breathe. The crust is for all practical purposes solid oxygen with a few impurities. Imagine a crooked entrepreneur advertising oxygen at a bargain price and then simply delivering the bewildered customer a box of rock with a label prominently displayed that certifies the contents to be oxygen 94 percent, preservatives and inert ingredients 6 percent.

The idea of preservatives and inert ingredients is more than a confidence man's trick. The major elements in rocks and minerals such as silicon and aluminum are strongly bonded to the oxygen atoms by forces called chemical bonds that in effect glue the oxygens to one another to form a solid crystalline structure. Without the "preservatives" the oxygens would combine in pairs and float around in the atmosphere. There are a few elements that do not readily combine with oxygen. They exist in rocks as pure elements or in combination with one another as "inert ingredients" in the sense that they play no role in sticking the oxygen atoms to one another.

Thus most of the minerals that make up crustal rocks consist of oxygen atoms packed together in orderly arrays that approach the closest possible packing of spheres. Such a grouping of roughly spherical atoms leaves small interstices, or spaces, that are occupied by the other atoms that are component parts of the minerals. Size relationships are such that a silicon atom fits nicely in the small hole between four oxygen atoms packed in such a way that each one touches all the others. Magnesium and the kind of iron that combines with one oxygen to form FeO fit into the space surrounded by six oxygens. Aluminum is intermediate in size and will fit in either kind of site, as will the other type of iron, the kind that combines with oxygen to form Fe_2O_3. Calcium and sodium are still larger, and each fits into a grouping of eight oxygens arranged as if they were at the corners of a cube, and the very large potassium requires an arrangement of twelve oxygens to create a hole big enough to accommodate its massive girth. Most minerals in the crust are silicates— arrays of oxygen and silicon plus different amounts of other elements to fill the remaining voids.

Other minerals include the compounds called carbonates that consist of a basic oxygen structure made up of three oxygens that touch one another and a carbon in the tiny space between. Carbonates comprise the abundant limestones and dolomites of the sedimentary rocks of the crust and are also a common minor mineral in many rock types. The hard parts of snails, clams, oysters, corals, and a myriad of invertebrates are made of carbonates, as is much of the mineral matter secreted by many plants, especially the one-celled algae.

Vertebrates and arthropods (insects, lobsters, and crabs) build their skeletons out of another oxygen compound, this one consisting of phosphorus in a site surrounded by four oxygens much as is the case with silicon. The mineral apatite also contains calcium and hydrogen in other sites. The hydrogen is paired with an oxygen to make a combined entity called hydroxyl, introduced earlier as the tightly bound water in a chemical analysis of a rock or mineral. The basic apatite structure is able to accommodate lots of other elements as well, and natural apatites thus have a lot of chemical variability. It is intriguing to note that the name *apatite* comes from the Greek word for trickery or deceit, which makes you wonder if the Greeks knew of its chemical trickery. I'll discuss some applications of the variability of apatites in vertebrate skeletons in a couple of the other chapters in this volume.

The only common elements that are as large as oxygen are fluorine, chlorine, and sulfur. All these elements form minerals that are analogous to silicates in that the large atoms are packed together with other, smaller ones occupying the interstices.

Geochemists and mineralogists tend to lump together elements that have similar chemical properties and sizes and speak of them as substituting for one another in crystal structures. This concept was formalized many years ago by V. M. Goldschmidt. Size is the more important factor, which means that in crystals chemically dissimilar elements like calcium and sodium replace one another readily because they are both about 1 Angstrom unit in radius. (An Angstrom unit is 0.00000001 centimeters, and there are about 2½ centimeters in an inch, so it would take a quarter of a billion of them to equal an inch.) A rarer element, strontium, can replace calcium both by virtue of size and chemical properties, a fact made use of in studies of bone apatite, as will be discussed in Chapter 10.

Because of the ability of one kind of atom to replace another kind in crystals, there is a rather limited number of common mineral types in the crust and in living things. In other words, each combination of elements forms, not a unique mineral or crystal structure, but a variant of a common mineral type. To be sure, there are over two thousand known minerals, but over 90 percent of them are rare.

The situation calls to mind an orchestra in which the conductor can replace one violinist with another of similar musical ability without changing the overall character of the orchestra very much. It doesn't matter whether the substitute is white or black, male or female, Hungarian or Mexican, as long as the violinist is of comparable competence to the one being replaced. Taken as a whole, the orchestra is the same, since all the chairs are occupied by the right sorts of musicians.

The earth's crust is composed mostly of a mineral called feldspar, which is made of tightly packed oxygens with silicon and aluminum and some combination of calcium, sodium, and potassium in the holes. Second in abundance is quartz, which is oxygen plus silicon. There are also three common types of minerals that contain iron and magnesium as important components, a little phosphorus-containing apatite, and that is about it. All the remaining elements are either hidden away in the major minerals—the so-called rock-forming minerals—or form very rare minerals of their own. The rare minerals are the stuff of collectors' dreams, the treasure of prospectors, and the bane of a student in a mineralogy laboratory course, but they are just that—rare. The crust is for the most part simple in composition and in the crystal structure of its component minerals. It is the departures from this overall simplicity that are the stuff of geologic research; it is the differences that make all the difference.

Chapter 2

The Age of Rock

*B*EFORE YOU READ WHAT I HAVE TO SAY ABOUT THE AGE OF rocks and of the earth I suggest that you try to put yourself in the place of the people who first considered this question. I recommend this exercise because it may perhaps give you a sense of the difficulty of the question, and in so doing make you a bit more sympathetic to the seemingly silly or stupid estimates made by scholars in past times.

So go back in time and imagine that you have walked out into the countryside to be alone and think. Your associates think that you're a little off in the head for always pondering irrelevant questions, so you have sought solitude to consider the age of things in the natural world. You know your own age because your mother told you, your family has a history that carries back for a few hundred years, and mythology accounts for earlier times, but what about the age of the

For more information, see references 1, 10, 21, 22, 27, 35, 36, 41, 43, 47, 56, 65, 80–84, and 89 in *Further Reading*.

objects around you? Perhaps you have found some stumps of very old trees and discovered that their age can be told from the number of growth rings in the wood of the trunk. But what about the rocks and the oceans and the stars in the sky? Are they 1,000 years old? 10,000? 100,000? How could you find out? Perhaps you have speculated about the very nature of time, and the notion of eternity might have occurred to you, at least in a forward direction. Does eternity extend backward too? Have the earth and the universe existed forever? These tough questions have puzzled scholars of disparate leanings for uncountable centuries.

One of the problems that we all have with time is the same problem we have with money—a little bit of either is easy to comprehend, but a large amount is so far beyond our frame of reference that intuition is of little help. A banker may feel comfortable dealing with someone else's billions of dollars as a basic unit, and a geologist may think of twenty million years ago as comparatively recent, but both these specialists have simply gotten used to the scale of the numbers that they use daily in their vocations. They don't necessarily have a better feel for large numbers than does anyone else. Without preconditioning, the idea of a few hundred years is a long time, and a few thousand verges on the mind-boggling. Little wonder, then, that thoughtful people in medieval times were content with the Christian dogma that the age of the earth and indeed of the universe was about six millennia, as determined from the Bible and other old sources. One Irishman, an Archbishop Ussher, made his place in history by doing an especially careful study of old texts and computing that the earth was created in 4004 B.C., on the 26th of October at 9:00 A.M. (presumably Greenwich mean time), attaining an accuracy unparalleled by even the best modern methods, sacred or secular. It is important to recognize that while some of us may poke fun at Bishop Ussher, his determination of the beginning of things was accepted as fact at the time, even by such scientific luminaries as Isaac Newton, and is still considered correct by huge numbers of believers in the biblical version of the earth's history.

An early secular attempt at determining the age of the

earth was suggested by the seventeenth-century English astronomer Edmund Halley of comet fame. His idea, which seems remarkably naive to us, was to measure the salinity of the ocean and then to measure it again ten years later. He thought that the amount of dissolved solids would have increased after a decade and that one could then use the rate of increase to project backward to a freshwater ocean, resulting in a determination of the age of the oceans and hence the earth. Halley's assumption that the seas were growing saltier with time was an incorrect one, but even if it had been right the method was doomed to failure. Assuming that the oceans are about three billion years old, the increase in salinity over ten years would have been only 0.00003 percent, a difference so small as to be undetectable by even the most skilled analyst. Furthermore, normal variation in open-ocean salinity is 100,000 times larger than the variation he could have expected in a decade. Of course, Halley was in the cultural milieu of Ussher's six-thousand-year-old earth, so we must forgive him his misjudgment. It is not known whether Halley actually carried out his proposal.

An attempt to actually calculate the age of the ocean from its salt content was tried toward the end of the seventeenth century by the Irish scientist John Joly, a professor at Dublin, who assumed like Halley that the ocean was initially fresh and that it accumulated dissolved salts from rivers that carried the same amount of solids in solution as present-day rivers. Joly allowed for a limited amount of recycling of sodium chloride from the sea back to the land. He came up with an age of the earth just short of 100 million years, a figure that was quite acceptable in 1899 to many scientists, although it was pointed out by disbelievers that the marine fossils from the Silurian period looked a lot like much more recent ones, suggesting that the salinity of the oceans couldn't have changed by a great deal in the interval.

Essentially the same approach was used by the renowned geochemist V. M. Goldschmidt as recently as the 1930s and 40s. Goldschmidt corrected for a larger amount of return of sodium to the land yet still came up with a result of 200 million years as the age of the oceans—still far short of the mark. We will look at the fatal flaw in his reasoning in Chapter 5.

In 1785 James Hutton, the great Scottish pioneer of modern geology, reported to the Royal Society of Edinburgh a widely quoted statement: "In the economy of the world, we find no vestige of a beginning, no prospect of an end." Hutton admitted that he didn't have a firm grasp on the exact magnitude of geologic time in that statement, which was a thinly disguised bit of praise for the theory of uniformitarianism, the concept that present-day processes were adequate to explain all events that took place in past earth history, events that must have been repeating themselves over and over again since a time very far back and would continue to do so in the future. It is supposed by some that Hutton imagined that the earth was infinitely old, but in the same paper he states that the ". . . periods of existence [stages in the earth's history] . . . [were] . . . in our measurement of time, a thing of indefinite duration," so it is clear that he meant that the earth was very old in human terms, but not infinitely so.

Hutton's concept that the earth proceeded in cycles, with no necessity for a beginning, was reinforced by Charles Lyell, who espoused uniformitarianism in his three-volume book *Principles of Geology*, which he began to publish in the 1830s, and in subsequent publications. A Mosaic chronology of prehuman earth history (i.e., as described in *Genesis*) had already been rejected by a majority of scientists, including many non-Huttonians. However, the doctrine of uniformitarianism existed partly to counter a religious view that the earth was formed by a succession of catastrophic events and miracles rather than by ordinary physical processes that took place gradually. According to catastrophists, vast amounts of time were not needed to explain the formation of observed natural features because they were created during brief events of great turmoil. Lyell recognized that the geologic processes that he observed were very slow, and while he then made no estimates of the length of geologic time, it was clear that he thought that ancient times were necessarily nearly unlimited.

Those were stirring times in earth science, and Charles

Darwin did little to dampen the enthusiasm when he published *On the Origin of Species* in 1869. Not that the ideas of organic evolution and natural selection were invented by Darwin. The philosopher Empedocles had stated in the fifth century B.C. that some creatures are less fit than others and that only the most fit would survive. The concept had been vigorously promoted by the biologist Lamarck and by Darwin's own grandfather, Erasmus Darwin. Charles Darwin, though, was able to integrate these concepts with those of Hutton and Lyell and to present convincing arguments that revolutionized the thinking of scientist and lay person alike.

Charles Darwin was a first-rate geologist as well as a biologist, and he had consumed Lyell's book during the dull, early part of the voyage of HMS *Beagle*. He stated in the first edition of *On the Origin of Species*, "He who can read Sir Charles Lyell's grand work on the Principles of Geology and yet does not admit how incomprehensively vast have been the past periods of time, may at once close this volume." In his book, Darwin made an estimate that the amount of time that had elapsed since the end of the Mesozoic era—the end of the age of dinosaurs if you like—was 300 million years. This figure would be high by modern standards (65 million years is more like it) but was at least in the right ball park. Darwin's reasonably good guess was made on the basis of his view of the rate of processes of erosion (removal of weathered rock) in a part of southeastern England.

Lyell had done much the same thing in 1867. He had the gut reaction, apparently purely an estimate, that a molluscan species would require 20 million years for a complete evolutionary change to take place, and since there were twelve such changes since the start of the Ordovician period (505 million years ago, according to a modern time scale) he placed the age of the beginning of the Ordovician period at 240 million years. This was a *very* good guess, but had no solid basis.

Lyell and Darwin had one big obstacle to convincing many of their contemporaries of their conclusion that the earth was very old. He was named Lord Kelvin, and he was perhaps the leading physicist of the times as well as a pillar of the church. He was not a believer in organic evolution, seemingly finding it at odds with his view of an orderly world governed by a creator, which he saw confirmed in the order and predictability of the physics of the times. Kelvin calculated that the earth and sun were much younger than the figures guessed at by geologists and biologists. Such matters were of considerable public interest at the time, as indicated by the sizable sales of Lyell's and Darwin's books. (Lyell's *Principles of Geology*, which had eleven editions, gave him enough income that he only held a salaried position for two years as professor at King's College.) Kelvin published his first work on the age of the sun and earth in a popular magazine of the time, *Macmillans Magazine*. Much of the public, including many scientists, held to the belief that the earth had been created through a series of catastrophic events of which Noah's flood was the last, and Kelvin was of this camp.

In good physicist style, Kelvin made the simple assumption that the earth had once been hot, presumably molten, and that it was cooling off. The rate of heat loss was something that could be measured in mines, caverns, and drill holes. Projecting backward from known heat-flow figures and present temperatures, Kelvin concluded, in a paper read to the Royal Society of Edinburgh in 1862 and published in the Society's transactions two years later, that the age of the solid crust was between 20 and 400 million years. In 1865 he read a paper at Edinburgh titled "The Doctrine of Uniformity in Geology briefly refuted" in which he sought to discredit Lyell's belief in an infinitely old earth.

Kelvin also assumed that the sun was cooling off through the loss of enormous amounts of radiant energy at a rate that could be measured easily. In 1897 he calculated that the sun's radiation would have been so much more intense that life could not possibly have existed on the surface of the earth until the last twenty to forty million years because the surface would have been too hot. He therefore preferred the shorter of his earlier estimates. Indeed, Kelvin is said to have considered perfectly acceptable a late–nineteenth-century estimate made by Clarence King, founder of the U.S. Geological Survey, that the age of the crust was twenty-four million years.

However elegant his assumptions and calculations when compared with the educated hunches of Darwin and Lyell, Kelvin lacked one vital piece of information: The heat flowing from the interior of the earth was not residual from a molten state but arose from radioactivity.

Antoine Henri Becquerel's discovery of radioactivity in 1896 was followed by the Curies' observation in 1903 that radium compounds gave off heat. T. C. Chamberlin of the University of Chicago had already guessed by 1899 that elements contained a vast reserve of energy. Prior to the Curies' discovery that heat was generated by radium, Ernest Rutherford and Frederick Sody, then at McGill University, discovered that energy was released in the emission of particles called alpha activity by radioactive elements, and they had concluded by 1904 that all radioactive elements contained great latent energy. With these new observations, the concept that the earth was originally hot and was gradually cooling off could be abandoned in favor of the theory that the planet had an internal heat source that had been more or less constant for a long time. Nobody understood the actual mechanism for the sun's heat source, but it was assumed to be somehow related to the phenomenon of radioactivity which meant that the surface of the earth need not have been terribly hot twenty to forty million years ago. So much for nineteenth-century physics.

In the 1770s a surveyor named William Smith was engaged to lay out and supervise the building of a canal in England. His job gave him the opportunity and the time to observe rocks freshly exposed in the canal cuts. Smith, who had collected fossils as a hobby when a young man, soon noticed that the rocks, though tilted at present, were stacked in a regular succession and that each type of layer contained characteristic fossils. Intrigued, he began to collect data whenever and wherever he could. He was not one of the educated gentlemen scholars of the times and did not feel comfortable writing down the results of his efforts, but he conceived the idea of showing the distribution of the various rock units on a map, much as was done with the soils maps with which he had become familiar in his work.

Smith completed his work on the canal project around the turn of the century and then spent the following years seeking out surveying jobs that took him to parts of England and Wales where he could moonlight at geologic mapping. His now-famous map was published at his own expense in 1815 and did not receive immediate recognition by the community of dilettante scholars of the day, partly because one of their own, a Rev. Townsend to whom Smith had referred in a letter as one of "his first two pupils," had his own, poorer map published in 1813. In time, however, Smith's accomplishment was acknowledged for the great work it was, and he was even awarded a medal and other honors by the gentlemen who had once scorned the unlettered surveyor.

Smith had not only made the first real geologic map of a substantial area but had laid the groundwork for what is now called stratigraphic correlation, the practice of comparing the assemblages of fossils in one layer with those of another layer in some other location to demonstrate that they were the same layer. (As I'll discuss in Chapter 6, this is not the same as showing that the two layers are of the same geologic age.)

Correlation is important to geologists because there is no single place on earth where an entire, uninterrupted succession of layered rocks—from oldest on the bottom to youngest on the top—is to be found. Therefore, as Lyell taught, bits and pieces from diverse locations have to be correlated, and the *total* sequence exists only in the minds of geologists. Correlation isn't quite as simple as Smith and his contemporaries had thought, but it is fair to say that by the middle of the nineteenth century the stratigraphic picture of the fossil-bearing rocks that overlie (and hence are younger than) the ancient granites and their kindred was pretty well established. This being the case, the total thickness of sedimentary rocks that had been deposited since Precambrian time—the first major episode of earth history, which ended about 590 million years ago—was known. Thus there was a quantity that could be used to estimate the age of the oldest fossiliferous rocks of the earth (fossils are found in some Precambrian rocks, but they are rare and were not recognized during the nineteenth century).

Only one other number was needed: the rate of accumulation of sediments. Divide that rate into the total thickness of sedimentary rocks, and you get the age of the earliest rocks in the stack. The difficulty is that determining the rate of accumulation is easier said than done. This problem did not prevent earth scientists from going right ahead and making calculations based on shaky data, and between 1860 and 1909 eighteen estimates of the age of the earliest sediments were made by distinguished geologists that ranged from 3 million to almost 1.6 billion years, a spread that gives little comfort to the seeker of truth. It is worth adding that some geologists changed their estimates by as much as eightfold during the half-century interval.

The thickness-of-sediments method is beset with problems, the worst of which is that the rates of accumulation of different sedimentary rock types are altogether different from one another and are different under different circumstances. A foot of sand might accumulate in a few minutes whereas a foot of fine mud might take one hundred years or longer. Even within a single sediment type the variation is immense. A layer deposited particle by particle might take a very long time to accumulate, while a similar-appearing layer could be deposited almost instantly by a slide or slumping mechanism. In other places sediment is removed rather than deposited, and the rate of removal can be slow or fast. The only way to obtain a reliable figure for average rates of sedimentation is to know the age of the top and bottom of the thickness and calculate it. In other words, one must already know the answer that is being sought—the age of rocks.

Thus at the end of the nineteenth century, earth scientists were drifting aimlessly in their search for a less subjective way of finding the age of the earth than crude guesses about rates of biological change or sedimentation. They knew intuitively that Lord Kelvin was wrong, but they couldn't prove it. Then along came the destruction of Kelvin's calculations by the discovery of radioactivity as a source of heat, and the question of the earth's age was open again.

Physical scientists vigorously pursued the phenomenon of radioactivity, and radioactive forms of the elements potassium, rubidium, and thorium were added to the list of radioactive elements, which already included uranium and radium. Physicists Rutherford and Sody proposed in 1902 that radioactivity was a side effect of the spontaneous transformation of one element to another and showed that helium was one product of the breakdown of uranium. By 1906 Rutherford was using the ratio of uranium to helium in minerals to calculate the age of formation of the minerals. Any helium trapped in the crystal structure must have formed from the decay of uranium, so the more helium, the older the crystal. He was able to measure the rate of the breakdown of uranium experimentally because the so-called alpha particles which are emitted through radioactive decay are actually helium nuclei. This alpha activity can be counted to provide a measure of the decay rate.

The uranium-helium method of dating a mineral depended on one major assumption that brought the question of the age of the earth full circle back to Hutton and Lyell. The assumption that Rutherford and others readily accepted was that the rate of radioactive decay was the same in the past as it is now; in other words, they assumed uniformitarianism. Thus using radioactivity for estimating the age of rocks did little or nothing to change the minds of those whose understanding of earth history came from the Bible, nor has it to this day. There were also scientific objections that temperature or chemical reactions might affect the rate of radioactive breakdown, but these criticisms have been found through experiment to be without merit.

In 1905 Yale chemist B. B. Boltwood brought American science into the act by proposing that lead was also a decay product of uranium, and in 1907 he showed that in a given mineral the lead:uranium ratio increased with age. He used rough estimates of the rate of radiogenic (i.e., arising from radioactive decay) lead formation from uranium to calculate the age of a number of mineral samples. His results are in remarkably good agreement with accepted modern values; in fact, when one considers the analytical techniques with which Boltwood had to work, his results are nothing short of miraculous. Radioactivity had provided earth scientists with an objective way to measure time.

The principle of radiometric dating is that a given radio-active element will break down to one or more "daughter" products at a constant rate. The rate of radioactive decay is generally expressed as the length of time it takes for half of the original amount of the radioactive element to break down—the so-called *half life*. In theory, when the first half life is over, the remaining quantity will take the same amount of time to be half gone, and that remainder the same length of time and so on, with the result that the total decay process takes infinitely long. To understand this concept better, suppose that you are traveling from New York City to Buffalo, a distance of roughly 440 miles, by going in a series of steps, with each step going only half the distance. Theoretically, you will never really get to Buffalo, only closer to it, because after the first half of the distance is traveled you will still have 220 miles to go, then 110, then 55, then 27.5 and so on. With the remaining leg becoming smaller with each step until it becomes infinitesimally small. This is wishful think-ing, of course, because you will get to Buffalo in real life, and in real life radioactive elements with relatively short half lives—for some a matter of just a few seconds, days, or years—are present in the crust in such small amounts that for all practical purposes they are absent.

In order to explain radioactivity, I'll have to discuss the ele-ments a bit more specifically. An atom of an element consists of a central, dense core, or nucleus, made up of positively charged particles called protons and neutrally charged ones called neutrons. The nucleus is surrounded by less massive, negatively charged particles called electrons. In chemistry, elements are classified by the number of protons in their cores, the so-called atomic number of the element, ranging from 1 for hydrogen to 92 for uranium among the common, naturally occurring elements. The mass number, or atomic mass, is the sum of the number of protons plus neutrons in the nucleus. Atoms with the same atomic number but different atomic mass are called isotopes of the element. Thus hydrogen, with atomic number 1, can have zero, one, or two neutrons in its nucleus, so there are hydrogen atoms with mass numbers of 1, 2, or 3, which would be symbolized as H^1, H^2, and H^3 and even have different names—hydrogen, deuterium, and tritium, respectively. These are the isotopes of hydrogen. Tritium is radioactive (i.e., capable of emitting radiant energy as a result of the spontaneous disintegration of its nucleus) and therefore called a radioisotope.

The principal radioisotopes that are used for the radiometric dating of minerals and rocks are potassium^{40}, rubidium^{87}, uranium^{235} and uranium^{238}, all of which have half lives of about one billion years or more.

Potassium, whose chemical symbol is K, is one of the more abundant elements in the crust. It is present in practically all igneous rocks and many metamorphic and sedimentary rocks as well. Most potassium in the crust (93.08%) is the ordinary, nonradioactive kind, K^{39}; another 6.91 percent is a nonradioactive isotope, K^{41}; and 0.0119 percent is the radioisotope K^{40}. Potassium^{40} breaks down into calcium^{40} and the gas argon^{40} in a fixed ratio. Calcium^{40} is common calcium (symbol Ca). As such it is useless for radiometry because calcium is present as a major constituent of most rocks and minerals, and the radiogenic kind cannot be distinguished from the ordinary kind.

Argon^{40} is the common kind of argon (chemical symbol A) present in the atmosphere owing its origin to the decay of radioactive potassium over billions of years, so you might think that A^{40} is just as useless for radiometric dating as Ca^{40}. Argon, however, is one of the noble or inert gases, so named because they do not combine to form chemical compounds except under very special laboratory conditions. Any argon found in a crystal structure must therefore have arisen from the radioactive decay of potassium because it could not have become part of the original crystal at the time of its formation. Thus the radiometric dating of potassium-bearing minerals is possible by measuring the levels of the appropriate potassium and argon isotopes in what is called the K/A method.

Using argon for radiometric dating does present some problems, though. Because it is a gas, argon can leak out of crystals if they are damaged in any way. Atmospheric argon can also contaminate the crystal by being sorbed onto its surface or even by entering the mineral through the same sort of cracks and other defects that might let gas escape. These difficulties were not, however, enough to discourage earth scientists and physicists because K^{40} is so widespread in rocks that the K/A method promised almost universal application. Also, the half life of K^{40} is about 1.3 billion years, which meant that scientists might be able to date the oldest rocks on earth—Boltwood's oldest were estimated to be about 1.6 billion years old. Such a half life is still short enough that dating very young rocks would also be possible, perhaps even rocks only a few million years old or even younger. The promise of this method was too great for scientists to give up in the face of difficulties.

Determining the content of K^{40} is easy. The total potassium is measured by ordinary chemical analysis, and the amount of K^{40} is calculated using the well-established proportion of K^{40} in naturally occurring potassium in the earth's crust.

Determining A^{40} is not as simple. The instrument used to do so is called a mass spectrograph. It works by shooting atoms out of a source gun into an evacuated (vacuum) chamber. These fast-moving atoms are subjected to electrostatic or magnetic forces that deflect them proportionally to their mass so that each type of atom, including argon, can be counted separately from the others present. This sounds simple enough except that mass spectrographs were all made of metal in the 1950s, and metals have the property of absorbing gases like a linen napkin absorbing beet juice. Once a metal has absorbed gas atoms they are as tricky to get out as that stain in the serviette. Thus metal spectrographs could be used to make age determinations of very old samples for which the amount of argon was substantial but were ill suited for dating young minerals and rocks because contamination of the spectrograph by atmospheric argon was too severe compared with the tiny amount of argon in the sample. To the rescue came J. H. Reynolds, an experimental physicist at Berkeley, who made an all-glass spectrograph that could be cleansed of contaminating argon so that really tiny amounts from a sample could be measured with precision and accuracy.

Garniss Curtis and Jack Evernden of the Berkeley geology and geophysics department were campus neighbors of Reynolds in the 1950s and quickly undertook to use the glass spectrograph for dating young rocks with an unprecedented degree of accuracy. They first turned their attention to dating the early–middle-aged rocks of the Sierra Nevada and Coast Ranges of California and other regions that had been formed in the tens to the low hundreds of millions of years ago. Their data supported the concept that there had been movements of the crust for hundreds of miles along the San Andreas fault and added fuel to what was then a hot controversy, although it is essentially universally accepted now (at this time many conservative geologists did not believe that there had been movements along the San Andreas fault for more than a few tens of miles).

The challenge of dating still younger rocks was irresistible, so Curtis and Evernden tried to close the gap between K/A dating and the carbon[14] method. Carbon[14], a radioisotope of carbon (common carbon is mostly the stable isotope carbon[12]), is formed in the upper atmosphere from nitrogen and is then incorporated into plants and other living things. It can therefore be used for estimating the age of organisms. However, the half life of C^{14} is so short (5,750 years) that it is useless for any sample older than about 40,000 years.

Curtis and Evernden worked with volcanic rocks from Europe and Africa to test the chronology of the last few million years, and their work led naturally to interaction with anthropologists, notably the well-known Leakeys, who were studying ancient human and prehuman sites in Africa. At the sites where the Leakeys and others found their fossils, volcanic activity had left layered rocks that contained crystals and volcanic glass suitable for dating by the K/A method, and the lab at Berkeley hummed. The results were surprising to many who had imagined that man and his ancestors dated back only a few tens of thousands of years when in fact it seems that a few millions is more like it. The presumed antiquity of early human and prehuman sites was viewed with disdain at first, but now there is general acceptance of the facts revealed by the K/A method—Bishop Ussher notwithstanding.

Today a variety of methods are available for dating rocks, minerals, and other objects so there are ample cross-checks to convince all but the stubbornest traditionalists. It appears that the oldest rocks in the crust, found in Greenland, are about 3.8 billion years old. In Australia, a mineral called zircon has been dated as 4.2 billion years old. The zircon crystals are older than the rock that presently contains them, suggesting that they derived, by weathering, from a rock as old as they and then became part of a new sedimentary rock where they are now found. There are reasons to believe that the initial formation of the earth took place about 4.8 billion years ago, and the whole cosmos dates back 8 billion years if calculations based on a theoretical origin of the elements are correct. It is a useful insight into the scientific mind that neither Kelvin nor Joly *ever* admitted publicly that the discovery of radioactivity and its effects made any modifications necessary in their age determinations, and Joly even published a paper in 1923 that defended his low estimates.

It is well to remember that, prior to the discovery of radioactivity, the scientific community generally agreed with Lord Kelvin's grossly incorrect estimates of the age of the earth and sun, so neither lay person nor scientist should be so complacent as to think that now we know it all. As unlikely as it may seem, there just might be some fundamental discoveries made that could dramatically revise twentieth-century truth.

Chapter *3*

Warm Heart, Cold Sky

*I*N THE AREA WHERE I LIVE—RURAL MINNESOTA—MANY OF THE residents have a blurred view of heat flow in the earth. Minnesota has cold winters, and a concern of all of us residents is that the buried water pipes might freeze. Generally speaking, water in the pipes doesn't begin to solidify until the end of winter or even into spring. After making it through the long winter without frozen pipes, it is discouraging to have the problem crop up just when conditions above ground are getting pleasant. The locals explain this phenomenon by saying that "the warmth forces the cold down," showing clearly that they have little or no intuitive grasp of that fundamental law of thermodynamics that states that heat can only flow from hot to cold. According to the third law of thermodynamics, heat from the surface can't make cold move, for cold is only the lack of heat, and only heat can move. I

For more information, see references 11, 13, 56, and 89 in *Further Reading.*

don't think that it is strictly a lack of education in the sciences that makes the locals think that cold can move. In this land where gravediggers have to burn a wood fire over a gravesite for several days before a hole can be dug, cold is a very real thing, much too real, almost palpable, to be just the lack of something. It is not the lack of heat; it is *cold;* it can move and freeze pipes too. Cold has a smell, a sound; cold *is* something to a Minnesotan; in the words of one resident, "When I open the door, cold comes in—it wants to." In the folklore of Minnesota, the laws of thermodynamics have exceptions.

What, then, is the thermodynamicist's explanation of Minnesota's freezing pipes? Winter is a time when the heat from the first few meters of the earth is given off to the frigid atmosphere, which lowers the ground temperature to a level approaching the ambient air temperature at the surface. At any point in the ground, the temperature gradually drops and the freezing level moves downward, approaching the waiting water pipe, which, unlike earthworms, cannot move deeper to escape its icy grip. When spring approaches, the surface temperature begins to rise until there is a time of no net heat flow. Heat then begins to flow into the ground and surface dwellers think that the big thaw has started. Down at pipe level, however, the ground may have just reached the freezing point. A foot or two above the pipe the temperature is still well below freezing because the heat from the surface has yet to reach that depth. Therefore, at pipe level, heat is still flowing upward, dropping the temperature still further, and the pipe freezes. Until the heat working its way down from the upper layers reaches the buried pipe and melts the ice, it will stay frozen unless man intervenes by heating the pipe or digging it up, no treat with frozen ground.

Minnesotans notwithstanding, the remainder of the earth follows the rules of thermodynamics, and heat moves from hot places to cooler ones. There is a branch of geophysics whose practitioners study the distribution and movement of heat in an attempt to explain features of the hidden interior of the earth. The earth, while not an energetic object like the visible stars, is nevertheless warmer than the near absolute zero that characterizes space. (Absolute zero, or zero degrees Kelvin, at approximately $-273°C$ or $-459°F$, is the hypothetical point at which all motion in matter ceases, which is to say there is no heat energy in the matter.) In fact, as Lord Kelvin knew, if you measure temperatures in mines or boreholes, you'll find that they increase downward, indicating that there is a source of heat within the earth. Those of you who have visited mines or caves, or even a basement under a building, may take exception to the statement that temperatures increase downward because you know that it is cooler in the underground opening. Others with the same subterranean experience may find the notion that temperatures increase downward perfectly acceptable. The difference depends, of course, on what season of the year you visited the underworld. In summer the cave is cooler; in winter it is warmer. Subsurface temperatures are generally stable throughout the year if the overlying rock and soil is thick enough to damp out the seasonal changes. The depth at which temperatures remain stable depends on the character of the overburden, principally on its ability to conduct heat, but in most places it is about 4 to 6 meters (thirteen to twenty feet). Below that depth, where surface influences vanish, the earth's temperature does increase downward.

Since the interior of the earth is hotter than the exterior, heat must flow outward. The rate of heat flow from the interior is a function of the thermal conductivity of the rock and the temperature gradient. The more conductive the rock, the more heat can flow through it. A temperature gradient is just a way of describing the difference in temperature between the hot and cold sides of a chunk of matter. The bigger the difference for a given distance and conductivity, the larger the gradient and the larger the heat flow from hot to cold.

The earth's geothermal gradient, or temperature increase with depth, ranges from about 8° to 40°C/kilometer (23° to 117°F/mile) in most parts of the crust on land and is a bit higher in deep sea sediments. You don't have to be a geophysicist to recognize that the temperature will reach the boiling point of water at a depth of only a few kilometers and that the deep interior of the earth must be very hot indeed. Some regions of the earth have much higher heat flow over limited areas of a few hundreds of square kilometers—for example in Yellowstone Park; The Geysers in California; Larderello, Italy; and Wairakei, New Zealand. The last three have even been exploited commercially as energy sources. Yellowstone has so far escaped such development. Such geothermal areas doubtless owe their unusually high heat flow, as much as a thousand times the average, to local heat sources such as a buried body of cooling igneous rock.

Rocks are poor conductors of heat, as you might suspect from the earth's rather high thermal gradient. Good heat conducting materials don't support much of a gradient, which you can prove to yourself by sticking a silver spoon into a cup of hot coffee. The spoon handle quickly gets too hot to touch because the thermal gradient is almost zero. In other words, the handle quickly reaches almost the same temperature as the coffee.

By contrast, the conductivity of the earth is about 0.004 to 0.010 calories/centimeter/second/°C. Somewhat meaningless by itself, this statement means that the heat flow averages only about 1.5 microcalories per square centimeter per second. In other words, only a little more than one-millionth of a calorie of heat passes through an area of 1 square centimeter (about ⅜ inch × ⅜ inch) in one second. A calorie is the amount of heat needed to raise a cubic centimeterr of water 1 degree Celsius. Put another way, if we had 1 liter (a little more than a quart) of water at room temperature (20°C) and we wanted to raise it to the boiling point (100°C) from the heat flow through 1 square centimeter of the crust it would take a bit over 2,500 years to do so (I'm assuming that none of the heat escapes from the water). Viewed in that way, the heat flowing from the interior of the earth to the surface seems almost nonexistent. Put another way, though, the total heat flow over the entire surface of the earth in one year is a staggering 2.4×10^{20} calories per year. (To convert to ordinary notation instead of exponential notation, move the decimal point twenty places to the right and you will get 240,000,000,000,000,000,000 calories per year.) That is enough heat to raise 30 million cubic kilometers (about 7.1 million cubic miles) of water from room temperature to the boiling point. On a worldwide basis, then, the amount of energy that flows outward is substantial, at least by human standards.

However large it seems, even the huge heat flow from the interior is trivial when compared with the solar energy the earth receives, which is on average about 5,400 times as great. The relative importance of solar and internal heat sources indicates that short-term changes of climate in the past, such as the relatively recent cycles of glaciation alternating with warmer episodes, cannot be explained on the basis of changes in heat flow from the interior inasmuch as surface temperatures are almost solely controlled by solar input. Put in terms of the water heating example, the solar input would raise thirty-eight billion cubic miles of water, roughly a hundred times more water than is in the oceans, from room temperature to the boiling point in a year. It's clear that the water example cannot give us a feeling for the amount of heat. The enormity of the energy is better grasped if you realize that in five hundred years it would melt the rock of the outer part of the earth to a depth of about a kilometer (0.6 mile).

None of these things happens, of course, because heat is not retained by the earth but is lost to space by radiation, just as heat is radiated by a hot wood stove. The laws of thermodynamics tell us that any body at a temperature above absolute zero will emit radiation. Thus the earth, with an average surface temperature of about 10°C, emits radiation, mostly in the infrared part of the spectrum, into space. (Infrared radiation is not visible to the human eye. It is a longer wavelength and hence a less energetic sort of radiant energy than the visible portion of the spectrum.) An imaginary object that absorbs every bit of radiation that strikes it, reflecting none, would emit exactly the same amount of radiation as received by the earth at an equilibrium temperature of 6°C. The fact that the average surface temperature of the earth is 10°C, close to the theoretical 6°C of an object that radiates away all radiation striking it, tells us that very little heat is retained—it is all radiated into space eventually.*

The actual picture is even more complicated since a great deal of solar radiation is absorbed by the atmosphere or clouds before it can reach the earth. Solar energy is temporarily stored in the form of vaporized water, in carbon compounds formed as the sun's radiant energy is converted to chemical energy by the photosynthesis of green plants, as free oxygen that has the potential to react with other materials to liberate heat, and so forth. Heat is also stored temporarily in the surface layers of the earth, with heat flowing into the solid earth during summer or on cloudless days and flowing back out again during winter and at night, especially when the sky is clear.

In terms of geologic work, the two heat sources, internal and solar, do quite different things. The flow of internal heat to the surface can be very high locally, as when a volcano or a lava flow brings large quantities of heat to the surface in the form of heated or molten rock. Deep in the earth, in that part called the mantle, heat is transferred more rapidly than would be possible by plain conduction. The heat moves by a process called convection wherein plastic, partly molten rock moves very slowly, with the hot portions rising and the cooler portions sinking. Hot mantle is less dense than cool mantle

---

*A black body that absorbs radiation is heated by the radiation. At any temperature greater than absolute zero, a body will emit radiation (this is not reflection). At 6 degrees the radiation re-emitted would exactly equal the radiation received by the earth. If more was retained than emitted, the body would increase in temperature like a black cat in a sunny window. The fact that the average surface temperature of the earth is 10 degrees, almost the same as the equilibrium temperature for a black body, tells us that the earth re-radiates almost all of the energy it receives rather quickly. The earth is not a perfect absorber or emitter, in fact it absorbs a little more efficiently than it emits so the temperature is a little above the theoretical 6 degrees. On a scale of −273°C to +10,000°C (or any other high temperature), six is very close to ten.

so it rises buoyantly, displacing cooler parts. Near the top of the mantle the heat is transferred to the crust by conduction and the mantle material, now cooler, sinks back into the main mass. This convective motion not only transfers heat more rapidly than conductive transfer, but it converts heat energy into mechanical energy, i.e., energy that can do work. The motion of the convecting mantle is transmitted to the crust as frictional drag and is the driving force behind all its movement—e.g., the motion of continents and large segments of the oceanic floor, the upward building of mountains, and the sinking of basins.

Thus while a certain amount of the heat that flows from the interior of the earth is lost directly to space, a great deal of it is stored in various forms. One form is storage in strain in rocks. When a rock is deformed, energy is stored in it either as permanent changes in the makeup of the rock or as elastic, recoverable strain. The latter strain is recovered in the form of earthquakes. The former is recovered as heat when the rocks react by weathering or some other process. Rocks or minerals weather in part by reacting with water or other materials at the surface to form a new set of minerals, releasing heat in proportion to the energy stored in the weathering rock.

Heat is also stored mechanically in the form of mountains and basins. A grain of rock at the top of a mountain has potential energy because it can fall, releasing its energy of motion in the form of frictional heat as it strikes other grains and interacts with running water. Similarly a particle on the ocean floor has potential energy because it can fall into a deeper part of the ocean formed by crustal movements originating from convective forces.

The solar energy contribution affects only the top few meters of the solid earth, so all the great geologic events— the great motions of crustal plates and the elevation of giant mountain chains, the building of volcanoes, and the rumbling of earthquakes—all have their driving power in the heat of the earth's interior.

As mentioned in Chapter 2, the main source of heat in the

earth is from radioactivity, especially from the radioactive forms of the elements potassium, thorium, and uranium. Potassium[40] was apparently the most important heat source during the first half of the earth's history but now only produces about a third as much heat as thorium and uranium. The gravitative pull of the moon and sun causes tidal deformation of both the solid earth and the hydrosphere, thereby accounting for some heat generation, but this heat is mostly dissipated as motion of the oceans rather than as heat in the solid earth. In the early history of the earth the separation of the earth into crust, mantle, and core released very large amounts of gravitative energy as heat. In fact, this early earth had substantially greater heat sources from conversion of gravitative energy to heat, the presence of more radioactive atoms than now, and substantial periodic deformation of the crust and outer, solid parts of the mantle resulting from closer proximity of the moon to the earth. The history of this early, hotter earth is recorded in the Precambrian rocks formed more than two billion years ago under conditions of elevated temperature that were never again reached in the crust.

The sun's energy has worked together with the earth's internal heat to form the active and changing planet we observe in the geologic record and still see today. The cycle of water from liquid to vapor to liquid again is the great transporter of rock from mountain tops to ocean basins. It is the driving force that built the vast thicknesses of sedimentary rocks that contain the record of life's evolution on earth. These sedimentary rocks also contain energy stored not only in the obvious forms of coal and petroleum but also in other materials capable of reacting to release heat. The great cycles of weather and climatic change can also be attributed to the effects of solar radiation.

Man has learned to make use of heat transfer in the upper part of the crust in a variety of ways, for instance, by using heat pumps to extract energy stored in groundwater to warm buildings in winter or by using the same groundwater as a heat sink to dispose of excess heat in the summer. Such high-technology approaches are fine, but there is a low-tech application of this very principle that has been used by the peoples of North Africa for untold centuries to make ice in the desert.

Deserts have properties other than high temperature that make ice manufacture possible. One is that the humidity is very low at all altitudes, which means that the air transmits radiation very readily with minimal absorption. Because water molecules absorb and re-emit infrared radiation, the low humidity also means that the atmosphere in the desert is "clearer" in that it does not retain or reflect infrared radiation back to the earth. Thus the desert floor has an unobstructed "view" of space with relatively minor interference from the atmosphere, especially in the infrared range, the energy range from which radiation is lost by the earth. Since the temperature of space is essentially zero, the movement of energy is all one way, that is, away from the desert floor. The Arabs take advantage of this situation by using shallow pans of water that are shielded from direct sun by a wall. During the daytime the water is not heated by direct solar radiation and is even cooled by evaporation. At night, the temperature of the air drops rapidly, especially so because of its dryness. The water loses heat through evaporation, conduction to the air, and direct radiation through the "window" of the clear desert atmosphere. The total heat transfer is enough to reduce the water temperature to below the freezing point, and the clever people of the desert have ice with no refrigerator. At least the desert people understand that heat moves from hot to cold, but then, their pipes don't freeze in springtime when the bluebirds return.

Chapter 4

Stranger in Paradise

M AN IS SUCH A SELF-IMPORTANT CREATURE THAT IT IS NO surprise that until the last several hundred years, scholar and peasant alike considered the earth to be the center of the universe. Astronomy put an end to that concept after some literally bloody arguments between astronomers and church officials—particularly during the first few centuries of the second millennium of the Christian era—and, alas, we must reluctantly admit that our planet is just one of a countless number rather than the unique place our egos would prefer it to be. One straw of uniqueness that the devoted anthropocentric can grasp is the highly unusual composition of our atmosphere, especially the presence of the oxygen so vital to most life on earth. To explain why we have oxygen to breathe and why the residents, real or imag-

For more information, see references 11, 13, 18, 25, 39, 45, and 67 in *Further Reading*.

inary, of other planets in our solar system don't, we have to look at the sequence of events that has taken place since the earth began.

There is no direct evidence for exactly how the world began, although there is a host of indirect clues that gives us a pretty good idea. It seems that all of the planets and other bodies in our solar system were formed at about the same time by the accretion of many fragments of material from a disk-shaped primeval cloud surrounding the sun. In the beginning, then, our earth was a three-dimensional mosaic of solid bodies widely ranging in size which were held together by their mutual gravitative attraction. Since temperatures in space are very low indeed—near the theoretical absolute zero—the solids comprising the newly accreted earth must have resembled present-day meteorites and would have included substances that, given present-day surface temperatues, we would normally consider to be gases. Our planet is still bombarded with space debris, as anyone watching shooting stars can tell you, but it is assumed that the accretionary process is slower today than it was four or five billion years ago because much of the material has been used to make planets.

As the mass of the earth increased, it began to heat up from several causes. According to Harvard geophysicist Francis Birch, the guru of those who study heat in the earth, the sources of the heat were, as mentioned in Chapter 3, the decay of radioactive elements and tidal friction, both of which were more significant heat sources five or six billion years ago than now. Birch calculated that enough heat would have been generated in the primitive earth after about 500 million years to melt iron. This iron was then able to migrate inward to form the core. The inward migration was accompanied by a release of gravitative energy, in the form of additional heat, which was sufficient to melt the mantle rocks partially, thereby forming a mantle that was divided into a lower layer with relatively little radioactivity and an upper, radioactive layer. The concentration of the radioactive heat sources in the upper mantle provided enough heat to melt off relatively low-melting-point silicate fractions out of the upper mantle, which rose and formed the crust. Birch proposed that internal temperatures reached a peak value of about 4,000° to 5,000°C about 4.5 billion years ago.

At that point there was a very hot earth whirling around in space, getting hotter until an equilibrium was reached between heat production and heat loss to space by radiation, much like the radiation from an infrared lamp or a hot wood stove. In such conditions the originally solid gases of the primeval earth would have vaporized to form an atmosphere. The early atmosphere wouldn't have lasted very long, though, for most of the gas atoms and molecules would have had sufficient velocity at the high ambient temperatures to escape the gravitational attraction of the earth and therefore were lost into space.

This is all very logical, but what evidence is there to support the contention that the earth lost its first atmosphere? The evidence comes from studying the chemical composition of the other planets in our solar system and of the cosmos as a whole. It is comparatively easy to measure the approximate chemical composition of distant bodies such as stars, and we have lots of evidence on the composition of our sister planets as well, from both earth-based observations and planetary probes. All the information shows that the earth's atmosphere is relatively deficient in elements that form gases at the broad range of temperatures presently found there.

Our atmosphere is deficient in argon, a relatively heavy gas, even though by virtue of its weight it cannot attain escape velocity under the present temperature conditions found on earth. Our atmosphere does contain some argon, but it is an isotopic form that is a product of the radioactive breakdown of potassium[40] and not a residue from the beginning of the earth's creation. We are forced to assume that either argon was never incorporated into the earth or that it was driven off at some later time by heating. Argon is such a ubiquitous element in the cosmos that it is very unlikely that the earth could have accreted without containing significant amounts of it. We are forced to accept the only other alternative: that it was driven off.

If argon were driven off, then much lighter gases such as nitrogen and oxygen would have been driven off as well. Any water vapor present would have suffered the same fate. Thus we are inevitably drawn to a picture of a hot earth with no surface waters or atmosphere. In other words, the present atmosphere must have formed much later in earth history, when temperatures at the surface were much closer to present-day values. We can also assume fairly that the gases in our modern atmosphere must have been derived from the solid earth because their ratios are not all like those of the sun or larger planets, which they would be had the atmosphere accreted from space.

Once the earth had cooled enough to sustain an atmosphere, what was the early composition of that atmosphere? Solid geologic evidence bears on this question. Perhaps the best evidence is contained in some ancient rocks in South Africa, the gold-bearing conglomerates of the Witwatersrand, which are dated radiometrically at about 2.8 to 2.6 billion years old. These are sedimentary rocks that contain rounded pebbles of quartz and other components. The roundness of the pieces indicates that they were transported by water and worn smooth in the process. Thus we can infer that surface temperatures were low enough to sustain a hydrosphere at the time, so it is equally reasonable to assume that there was an atmosphere. The atmosphere at the time of the Witwatersrand probably contained argon and nitrogen, with small amounts of water vapor and other elements and compounds including oxides of carbon such as carbon dioxide, carbon monoxide, and possibly ammonia, a molecule composed of one nitrogen and three hydrogens.

You will have noticed that I haven't mentioned oxygen. The Witwatersrand conglomerates tell us that free (uncombined) oxygen was absent by the presence of rounded pebbles of pyrite (FeS_2) and uraninite (UO_2). The mineral pebbles had been weathered from bedrock at some place and transported by running water to the site of their deposition in the gold-bearing conglomerates. This means that the pebbles would have been exposed to the atmosphere both at the site of their weathering from primary host rocks and during the long transportation process. Both pyrite and uraninite are highly unstable in the presence of oxygen, so we can conclude with confidence that the atmosphere was oxygen-free at the time. We also know that the Witwatersrand environment was not unique, for there are similar rocks of the same age at Blind River in Ontario. We may also safely assume that gases like carbon monoxide and ammonia were present in the early atmosphere because they are also unstable in contact with oxygen.

Other geologic evidence tells us that oxygen became a significant part of the atmosphere at a later time, about 1.8 to 2.0 billion years ago. Rock sequences presently exposed in India and around the Baltic that were deposited in late Precambrian time contain bright red strata. This red color is produced by the presence of a mineral called hematite, which has the composition Fe_2O_3. This form of iron oxide is more highly oxidated than the oxide FeO that predominates in igneous rocks. Its presence in the red sedimentary rocks indicates that the oxidating of iron had taken place under surface conditions. Hence there must have been free oxygen in the atmosphere. The stranger in paradise had arrived at last.

The evidence that oxygen began to accumulate in the hydrosphere and atmosphere about two billion years ago is reasonably good, but where did that oxygen come from? The only possible source is the solid earth. As I mentioned in Chapter 1, the outer parts of the earth are almost pure oxygen (94 percent by volume) so there was no shortage of supply. The problem is that the oxygen in crustal rocks is tightly bound to other elements to form stable crystalline materials that do not yield up their oxygen readily. Judging from rock analyses you might expect that a little of the oxygen could be liberated from rocks in the form of water, carbon dioxide, and carbon monoxide. Indeed, these gases are released from rocks heated in the laboratory and are important constituents of the vapors exhaled by volcanoes, although most volcanic gases are recycled surface waters. Thus, there was oxygen in a combined form in the early atmosphere, and all that was necessary was to separate it from its partners carbon and hydrogen.

Water is an attractive source of oxygen because the hydrogen-oxygen bond can be broken to yield free oxygen through a process called photochemical dissociation. The *photo* refers to light, and the ultraviolet portion of sunlight provides the energy needed to break the bonding of water. This process takes place in the upper atmosphere today. Even at the relatively low temperatures at the edge of the earth's gaseous envelope hydrogen attains a high velocity because of its very low mass and escapes into space, leaving the oxygen behind. How much the dissociation of water contributes to free oxygen levels is a subject of debate among specialists, but all agree that it would not account for all of the oxygen we observe now or infer from the geologic record. Another source of oxygen is the dissociation of carbon dioxide. Today this reaction takes place in green plants through the process called photosynthesis. There were no trees, grasses, or herbs to do the job in those distant times, but there is ample evidence that plants capable of photosynthesis were found in the seas about three billion years ago. These simple plants, called blue-green algae, leave characteristic accumulations of mineral matter called stromatolites, dome-shaped and layered bodies that have been found in rocks of many ages throughout the world, and there are even modern analogues of such algae. Furthermore, chemical compounds thought to have been derived from chlorophyll, the photosynthetic agent in green plants, have been found in the same three-billion-year-old rocks that contain the stromatolites.

This is all very well, but how did these relatively sophisticated plants suddenly appear? The answer is that they didn't just suddenly appear but were the product of long evolution from precursors that lacked the photosynthetic mechanism for utilizing the energy of the sun. In our oxygenated world it is easy to lose sight of the fact that many living things thrive in the absence of oxygen. These so-called anaerobic organisms make their living by deriving energy from chemical reactions that do not involve free oxygen. Such anaerobes make their home in our own intestines, among other places, doing such things as producing methane (CH_4) and hydrogen sulfide (H_2S) from a bowl of beans. Other types live inside jars of improperly canned foods and in incorrectly smoked fish, producing the deadly botulism toxin. Yet others live at the bottom of the Black Sea and in hot springs, causing the deposition of metallic sulfides and elemental sulfur from solution.

Inasmuch as we can state with assurance that the early atmosphere did not contain free oxygen, we may assume that the first life forms must have been anaerobes. There is not a physical or chemical niche today that is not occupied by one or more organisms, so it is altogether reasonable that the early earth with its anoxic (oxygen-less) atmosphere and oceans supported a complex community of organisms that got along just fine without respiring oxygen as long as the appropriate chemical energy source could be tapped—long before the first mutant learned the trick of catching a sunbeam and living a life of ease.

You may have noticed a minor discrepancy of a mere billion years or so between the first evidence of plants capable of photosynthesis and the appearance of free oxygen in the atmosphere. Not that a thousand million years is forever, but it is long enough to take seriously.

It is typical of the problems involved in unraveling earth history that the time factor looms in any explanation that scientists come up with. Many earth scientists fall into the trap of thinking that there is so much time to be accounted for that any process, no matter how slow and inefficient, is adequate to explain some record found in the rocks. In fact, this is not usually the case. The late Arthur Holmes of the University of Edinburgh (a pioneer of radiometric dating, who, as a student, had the good taste to switch from physics to geology) once warned his fellow geologists that they must not "take refuge in the immensity of geologic time."

The discrepancy between the appearance of photosynthesizing plants and the first evidence of oxygen in the atmosphere in significant quantities is an embarrassment of the opposite kind. What went on for a billion years between these two events? To resolve this apparent conflict we should look at the two processes that were at work during this rather lengthy gap: the formation of free oxygen and the consumption of free oxygen.

The addition of oxygen to the earth's atmosphere resulted first from the photochemical breakdown of water. This breakdown must have taken place throughout the atmosphere because there was as yet no ozone (three oxygens bonded together in an unstable molecule) to absorb the ultraviolet light in sunlight in the outer portions of the atmosphere. Once photosynthesis began to take effect at the beginning of the troublesome time gap, oxygen production must have increased, but by how much? We have to consider how much habitat was available to the blue-green algae. Ultraviolet radiation is deadly to most life forms so it is unlikely that any algae lived on land except in shaded nooks and crannies. Even in the ocean the algae would have had a limited range because they would have required some depth of water to protect them from ultraviolet rays. Yet they still required light to carry out photosynthesis, and as any skindiver will verify, light levels drop off rapidly with depth. So the algae must have been restricted to shallow water. We can say with confidence that the photosynthetic production of oxygen must have been much less than at present.

The other process affecting the buildup of atmospheric oxygen was its storage or consumption. Any oxygen formed by algae would have been absorbed by dissolving into the ocean and could not have entered the atmosphere in meaningful amounts until the ocean became saturated. Furthermore, in the Precambrian seas oxygen was an environmental pollutant of serious proportions. We sometimes forget that oxygen is a highly reactive element that must have been toxic to almost all living things in the early Precambrian seas. It is really just a waste product from the photosynthetic process of converting water and carbon dioxide into carbohydrates like cellulose and glucose, so even green plants must have gone through an evolutionary stage of developing mechanisms for tolerating the oxygen they produced. As the ocean waters became richer in oxygen, the anaerobic life community that had held sway for millions on millions of years must have diminished greatly as its habitat became ruined by the pollutant oxygen. The dead anaerobes would have themselves used up oxygen, reacting with it to form carbon dioxide and other compounds. Newly formed oxygen would have also reacted with carbon that had accumulated in the seafloor's sediments. The floor then comprised a considerable "sink" into which oxygen was lost.

Oxygen was also consumed in reactions with other dissolved gases such as hydrogen sulfide, carbon monoxide, and methane. The early, preoxygen ocean was rich in such gases, like a stinking swamp reeking of the odor of rotten eggs and marsh gas—the kind of water body we might today call polluted.

The atmosphere of the time contained the same unoxidated gases, so any oxygen evolving from the ocean would have been consumed in atmospheric reactions. At some point all of the reactive gases in both the ocean and atmosphere would have combined with oxygen and oxygen production would also have exceeded exhalation of volcanic gases. Oxygen would therefore have started to accumulate in the atmosphere.

There was, however, one more sink to absorb oxygen before any significant amount could have accumulated. You may recall that the iron found in igneous rocks is mostly in the state of oxidation that chemists call ferrous iron (a name given to iron possessing a positive charge of two that combines with oxygen to form ferrous oxide, FeO). The ferrous ion (an ion is a charged atom) is soluble in water, and the Precambrian oceans doubtless contained large quantities of dissolved iron in that form. As oxygen entered the ocean it would have reacted with the ferrous iron to form the insoluble compound ferric oxide, Fe_2O_3. There are worldwide deposits, many of them constituting the great commerical iron deposits of the earth of unusual, iron-rich rocks found in layers restricted to Precambrian rocks with ages of about two to three billion years, a period during which the oxygen produced was apparently absorbed in reactions with dissolved ferrous iron.

In 1972 geochemist Yuan-Hui Li drew up an estimated balance sheet for oxygen:

net oxygen produced by
    photosynthesis . . . .3.0×10^{16} tons
oxygen in atmosphere at present . . . . .0.1×10^{16} tons
oxygen used to oxidize ferrous iron . . .1.0×10^{16} tons
oxygen used to oxidize volcanic gases .1.9×10^{16} tons
    total . . . .3.0×10^{16} tons

If Li's estimates are the least bit accurate, then the billion-year gap is quite acceptable inasmuch as over 96 percent of the oxygen produced in the earth since the beginning of the atmosphere was consumed in reaction with ferrous iron or volcanic gases.

Once the little remainder of oxygen that was not consumed in reactions appeared in the atmosphere, the course of weathering changed markedly, with oxidation becoming a predominant mechanism. Anaerobic life was from that time on restricted to environments removed from the deadly, oxygen-containing air. The fossil record shows that life on earth expanded very rapidly in the last part of Precambrian time, with oxygen-breathing plants and animals now upon the earth. The power of the sun had been harnessed by life at last, and the world would never be the same.

Chapter 5

Salt of the Earth

O N A SCENIC MOUNTAINTOP IN THE AUSTRIAN ALPS THERE IS A small castle that is used as a meeting site for conferences of select groups of scientific specialists from around the world. The splendid isolation of Burg Wartenstein has provided a fertile ground for numerous meritorius scientific endeavors, including shelves of books arising from the efforts of the conferees. "The Castle," as it is known to in-group scientists, was not always a romantic retreat for great minds. It was originally built by an Austrian nobleman who was in the business of importing a utilitarian commodity, salt, from the Mediterranean. Salt was a valuable commodity in central Europe in those days, and the importer had to use some of his profits to construct a series of fortified villas to house the

For more information, see references 5, 6, 29, 38, 43, 46, 66, 68, 70, and 95 in *Further Reading.*

small private army that was required to keep hijackers away from his salt-hauling caravans.

Man has been interested in salt for noncommercial reasons as well: The question of why the ocean is salty whereas rain, rivers, and most lakes are composed of fresh water has intrigued many a scholar. Legend attributed the salty sea to the tears of an unhappy goddess, but that was avoiding the question. There is no direct evidence indicating whether the ocean was already salty when it was formed or whether it was fresh water that later acquired its salt from outside sources, but most investigators had little doubt that it was originally fresh because the primeval sea must have been formed by water condensing from the atmosphere as the earth's surface cooled to a temperature below the boiling point of water. With that assumption generally accepted, questions still remained: How old is the ocean, and where did the salt come from? To understand the salt content of the ocean, the geologist has to look at the mineral matter dissolved in it through the eyes of an accountant, balancing the inputs and outputs of salt to and from the ocean. A modern geochemist would call that a salt budget. The term *budget* is used here not as it was used by that long-ago Austrian entrepreneur but as a quantitative statement concerning how much salt resides in what places and how it moves from one part of the system to another.

In the case of salt—sodium chloride—there are really two budgets to consider, one for each element. The budget of sodium has received more study, so I'll deal with it here. The chlorine in the sea originated in emanations from volcanoes and the weathering of igneous rocks, but few details of its natural history are known with precision.

Geochemists have looked at sodium and its cycle for a long time. As you will recall from Chapter 2, the astronomer Halley and the geologists Joly and Goldschmidt all made the assumption that the ocean was initially fresh water and that it had become saltier with time. Halley had a sixteenth-century view of the age of the earth and devised an impractical experiment to determine its age from the sea's saltiness because he thought that the salt level would change enough in a few years to allow accurate projections back to the ocean's beginning. In 1899, Joly, still under the influence of Lord Kelvin's incorrect physical-theological estimate of the age of the earth, estimated the ocean's age at approximately 100 million years by using the salinities of modern rivers and assuming that the salt in the sea had been derived from similar rivers in the past. As late as 1937, and in his posthumously published book *Geochemistry* (1954), V. M. Goldschmidt still maintained that in principle the age of the ocean could be determined by a more sophisticated version of Joly's scheme, but even so came up with an age much too young by modern standards.

Both Joly and Goldschmidt were strongly influenced by the concept that the fundamental rocks of the crust were igneous rocks and that sedimentary and metamorphic rocks were secondary in importance because they were all derived from igneous precursors. Thus they presumed that the sodium in the ocean had its origin in the weathering of igneous rocks and that the amount of sodium residing in other rock types was insignificant. This attitude (also seen in Clarke and Washington's use of igneous rock analyses to estimate the composition of the crust) reflects the influence of a nineteenth-century German geologist named Rosenbusch, who viewed all geologic processes as a one-way street starting with igneous rocks, with weathering no exception. Hutton's concept of cycles, with old rocks being remade into new ones, had not caught on everywhere although it was popular with the French, who were inclined to disagree with the Germans as a matter of policy. The British and Americans took a mixed stand on Hutton's theory that reflected a certain amount of awe of the dogmatic Rosenbusch.

In 1960 Tom F. W. Barth, the distinguished Norwegian geologist, was president of The Geochemical Society, which met that fall in Denver. Barth, a tall, slender man with flowing gray hair and a commanding though gentle demeanor, stood at the lectern in the Hilton meeting room wearing his usual bow tie and gave a presidential address that took his late "countryman," Goldschmidt (many geologists think of Goldschmidt as a Norwegian, but he was actually a Swiss who was a favorite student of Professor Brögger's at Oslo and who spent most of his professional career in Norway) to task and changed the thinking of geochemists everywhere. Disdaining the influence of Rosenbusch on Goldschmidt and others, Barth minced no words: ". . . the dead hand of the past is still directing our thinking. . . ." Barth maintained that the sodium in the ocean is part of a cyclic movement of sodium in which large quantities are returned to the land masses by a variety of means and are then rechanneled to the sea again and again. Barth claimed that "the ocean is just a channel through which the migrating elements are cycled." This being the case, if the rate of sodium entry into the ocean were divided into the total amount there, as Joly and Goldschmidt had done, the result would give not the age of the ocean but the average residence time of a sodium atom in the oceanic part of the cycle. Barth's calculation of this residence time, 208 million years (the numerical equivalent of Joly's or Goldschmidt's age of the sea), is simply for one step in the cycle. Barth stated that his calculations showed that it would take about three cycles for equilibrium to be reached, so the *minimum* age of the oceans must be about 600 million years. This is a minimum figure because there is no direct way of determining how many cycles took place after equilibrium was reached.

In retrospect it is really amazing that it took sensible, intelligent people over two hundred years to figure out that geochemical budgets were parts of cycles, but it shows the power of mindset thinking. Barth emphasized that much of the sodium in the cycle was to be found in metamorphic rocks derived from sediments and concluded his address by stating that from the scanty data available in the literature of geology " . . . we have to conclude that practically all continental matter has been reworked, i.e., there is no 'juvenile' [original igneous] matter in the continents; all rocks have been sediments but modified and changed by metamorphism, . . . and remelting under plutonic [deep seated] conditions, then again raised to the surface to become sediments." Then Barth closed his speech by quoting Hutton who saw ". . . no traces [sic] of a beginning, no prospect of an end."

Barth had opened his address with a statement, "We have no quantitative knowledge of the composition of the earth as a whole, or the composition of the individual parts. The concept of geochemical cycles explains some of the migration of chemical species in the earth; but the quantitative relations of the cyclic processes are not known. Our ignorance is really shocking; and the underlying data are inadequate and no final solution would seem possible today." Whether you consider Barth a pessimist or a realist is your choice. There are certainly few data considering the magnitude of the system under study.

Nevertheless, it was only two years later that hydrologist D. A. Livingstone published a lot of new data on the sodium content of natural waters in a United States Geological Survey publication as well as a milestone paper on the sodium cycle and the age of the ocean. Livingstone cross-checked the validity of his sodium budget by seeing if all the various inputs and outputs concurred with the antiquity of the earth as determined by radiometric dating. He assumed from a variety of kinds of evidence, especially the fossil record, that the average rainfall and rate of chemical weathering of rocks on the land masses was not significantly greater or lesser than at present so he accepted present river transport figures for sodium as a reasonable guide to the past. From his data he calculated that 20.5×10^7 tons of dissolved sodium is carried to the sea each year by rivers. Of this amount, about 3 percent is assumed to be of industrial or other manmade origin, reducing the value to 19.9×10^7 tons.

The sodium in the sea does not, of course, all remain in the sea. A great deal of it is recycled back to the land by spray carried by the wind. Much of this spray is brought to the ground by rainfall. In a meteorological study of the United States, probably a sufficiently diverse country physiographically to be representative of the world's land masses, it was found that average rainwater contains about 0.6 ppm (parts per million) of sodium. Other studies have shown that rainwater is concentrated by evaporation and transpiration by plants by about three and a half times. Therefore the contribution of rainfall to river-borne sodium is about 2 ppm of the river-borne total. Sodium also returns to the ground directly as small particles, both as crystals and as bits of briny mist, at a level, as measured near Boston, of about a quarter of the amount of sodium in rainfall.

Livingstone concluded that of the sodium found in rivers no more than 3 ppm has been recycled through the atmosphere. Because the level of river sodium is only 6.4 ppm, sodium contributed by the atmosphere is a large amount, reducing the *net* (returned to the sea) river transport of sodium to only about 10.7×10^7 tons per year. If you were to assume that there was only a single cycle, as Joly and Goldschmidt did, then using this net transport figure 64.2×10^{15} tons of sodium would have been transported to the ocean since the start of the Cambrian period. (Livingstone used 600 million years for the start of the Cambrian period. Using a more modern value of 590 million would reduce the net transport figure to 63.1×10^{15} tons, but this difference is not relevant to the arguments that follow.) Yet there are at present only 14.1×10^{15} tons in the ocean (if it is fair to use the word *only* for a number as large as 14,100,000,000,000,000).

Thus the simplified budget is out of balance by a ridiculous 50.1×10^{15} tons. The missing sodium must presumably be sought in sedimentary rocks because there is no other place for it to have gone unless it has been lost to space or sucked down into the mantle—neither of which is a very likely hypothesis.

Some of Livingstone's numbers are of nongeochemical interest as well. Those of you who live in regions of the country where salt is lavished on the roads to melt snow and ice and to rust auto bodies will be suprised to learn that the amount of sodium contributed to the land masses by rainfall and solid particles is over forty times the total amount contributed by *all* industrial and other human uses. Unless the use of salt on roads has increased markedly in the past twenty years, mankind's salt contribution to the environment should have minimal or local effects only.

To verify the theory that the missing sodium ended up, or at least has been resident, in sedimentary rocks you need to know two things: the sodium content of a given unit of sediment and the total amount of sediments. Livingstone assumed that the sodium was sorbed onto the surface of sedimentary grains as well as dissolved in the water that filled the pores between grains. He also made the assumption that pore space amounts to about 20 percent of the entire sedimentary rock volume, an admittedly high value. Also, since pore waters in sediments are saltier than seawater he assumed that on average they have three times the salinity of the sea. He also assumed that about a third of the sorbing ability of sediments is used to hold sodium. All his assumptions were probably on the high side, and as we will see, it is useful to take this fact into account.

Using his high assumptions and also the highest estimates that other scientists had made of the volume of sediments on the sea floor, Livingstone arrived at a total estimate of the sodium in suboceanic sediments of 10.5×10^{15} tons. Reducing the 50.1×10^{15} ton discrepancy by that amount still leaves us with an unaccounted for balance of 39.6×10^{15} tons, no minor problem.

The missing sodium can only be in one place—in the sedimentary rocks that have been returned to the continents. These are not the same sedimentary rocks that are *now* on the continents because some of the rocks that were pushed up into the continental mountain chains and the like have been eroded away and returned to the sea again, perhaps several times. Basing calculations solely on the amount of sodium that is presently tied up in sedimentary rocks has been the fatal flaw in previous estimates of the age of the ocean. To quote Livingstone: "In effect, we have counted each sodium atom in our budget every time that it has been carried to the sea, but we have not counted it every time that it has been returned in uplifted sediments to the land."

Using estimates from a variety of sources, Livingstone concluded that the present sediments on the continents represent about one-quarter to one-sixth of the total volume deposited during geologic time. Two different geochemists—a
German, K. H. Wedepohl, and a Russian, A. B. Ronov—estimated that the total sedimentary rocks now on the continents
amount to about 800×10^6 km^3 (cubic kilometers), or about
5.1×10^{17} tons. Using Livingstone's correction factors, mentioned above, we could then estimate the total sediment
deposited as between six and nine times the present amount
or 20.4 and 30.6×10^{17} tons. We can use an average of 25.5
$\times 10^{17}$ of total sediment for some further calculations.

According to the late Columbia University geologist Arie
Poldervaart, sedimentary rocks contain on the average about
1.07 percent sodium. Applying this to the estimate for total
sediment above we could account for 27.3×10^{15} tons of
sodium, still leaving an unaccounted for deficit of 12.3×10^{15}
tons. We could try to resolve this problem in two ways. First,
we could assume that either the estimates of total sediment
or Livingstone's correction factors are wrong, but the values
were determined from a great deal of data so they are likely
reasonably accurate.

Second, we could assume that the value used for the
sodium content of sedimentary rock is wrong. Once again,
we must respect Poldervaart's estimate as being reasonably
accurate. However, his average included sedimentary rocks
laid down on land as well as those of marine origin. In fact,
nonmarine sedimentary rocks have much lower sodium content than their marine analogues.

In addition, by using the sodium content of average sedimentary rocks we are ignoring the fact that many sedimentary rocks are no longer sedimentary rocks at all but have
changed through the influence of heat and pressure and
chemical transformations into metamorphic rocks. In fact,
metamorphic rocks as a group contain more sodium than
sedimentary rocks, averaging 1.72 percent, according to
Poldervaart. However, many metamorphic rocks are derived
from igneous rocks and igneous rocks contain about 2.04

percent sodium, so an average for metamorphic rocks cannot be used to represent metasediments (metamorphosed equivalents of sedimentary rocks).

To account for the missing sodium in the budget, we need to establish either a higher average sodium content of sedimentary rocks or conclude that more of these rocks have been through the cycles. If we include some metamorphic rocks with the sediments, then we solve both our problems. Metamorphosed marine sediments contain more sodium than their unmetamorphosed equivalents because the sodium dissolved in the pore water becomes concentrated and, on metamorphism, reacts to become incorporated in solid minerals. The sodium content of metasediments would not be as high as Poldervaart's estimate because as mentioned, his average included metaigneous rocks as well, but it would still be appreciably higher than that of unaltered sediments. There is one additional matter that must be considered too. Many geologists believe that most of the granites of the Precambrian cores of the continents are really metamorphosed sediments and not igneous rocks at all. If we were to include the granites of such regions as metamorphosed sediments, then there is no problem at all in reaching a higher average sodium content for sediments and their derivatives. So we probably should use a higher average sodium value for sedimentary rocks and possibly increase the estimates of total sediments to include a large portion of metamorphic rocks and even a part of those incorrectly interpreted as igneous rocks.

The view that vast amounts of sediment have been subjected to metamorphism fits in very nicely with Barth's theory that the rocks of the continents have been through many cycles of destruction and metamorphism. Thus the missing sodium can be accounted for very nicely by having it tied up in sedimentary rocks and their metamorphosed equivalents. These sodium reservoirs do not all exist at present because we are considering all the sediment that has been returned to the continents during geologic time, not just what is there now. To summarize our salt (sodium) budget:

Carried by rivers to the sea . . . . 119.4 tons $\times$ 10^{15}
Returned by atmosphere to the land 55.2 tons $\times$ 10^{15}
    Net transfer to the sea . . . . 64.2 tons $\times$ 10^{15}
Dissolved in sea water . . . . 14.1 tons $\times$ 10^{15}
In suboceanic sediments . . . . 10.5 tons $\times$ 10^{15}
Returned in sediments to the land . . 39.6 tons $\times$ 10^{15}
    Net sodium in the cycle . . . . 64.2 tons $\times$ 10^{15}
    Unaccounted for balance . . . . . . zero

So we can reasonably account for the sodium in its cyclic path through sediments, down rivers, and through the air. The values used are incomplete and are estimates, but they are taken from various sources and were made by people with no axe to grind—at least not with respect to the overall budget of sodium. The fact that a combination of the various estimates gives an internally consistent picture means we can be somewhat confident in the outcome. Minor details of the picture will doubtless change as more data are accumulated by scientists in many diverse disciplines.

As I mentioned, Livingstone used the data to make a revised estimate of the age of the oceans. This activity had been given a bad name by Joly and a host of others but was not fundamentally wrong, just incorrectly executed in the past. Livingstone estimated the age of the ocean to be between 1.3 and 2.5 billion years, depending what assumptions he used. Such an age is still somewhat low compared with radiometric determinations, but it is in the right order of magnitude.

I have not expressly mentioned the deposits of rock salt that are found in many parts of the geologic record because their total quantity is small even although they are included in the calculations. Rock salt is a plastic material under conditions of burial and even reaches the surface of the earth in many places because it is less dense than ordinary rock so is able to "float" upward. The renowned Tabasco Sauce is manufactured on top of a rock salt deposit in Louisiana. Also, we are informed in the Bible that Lot's wife was turned into a pillar of rock salt for disobeying God. Doubters can travel to the region of ancient Sodom today and find rock salt bodies exposed at the surface, persisting because of the arid climate. The amount of naturally occurring rock salt is only 0.4×10^{15} tons, according to Livingstone. Since he wrote his paper huge amounts of rock salt have been discovered under the sediments on the bottom of the Mediterranean and off the northwest coast of Africa which would bring the total amount of rock salt to at least three times the old figure, or about 1,200,000,000,000,000 tons or more of rock salt. You could say that's a Lot.

Chapter *6*

Comparing Here with There

G EOLOGISTS' DESKS PROBABLY SPAN THE FULL RANGE OF neatness, from empty and dust-free to cluttered and seemingly disorganized. The same can be said for their tents, offices, suitcases, automobiles, or other parts of their working environment. I use the qualifying phrase—seemingly disorganized—because in the piles of papers, maps, reprints, books, letters, and other artifacts of the working geologist there is an inherent order that is uniquely understood by the members of the community of earth scientists. The order is imposed by the fact that objects in a pile generally accumulate by addition to the top. That is not to say that a given item might not have been tucked under a stack to hide it from prying eyes or that an item might not have been removed and replaced on top, but the succession of layers have an encompassing regularity that results in the oldest ones being on the

For more information, see references 1, 12, and 35 in *Further Reading*.

bottom and the youngest on the top. It is just this sort of order that is present in stratified rocks, rocks that accumulate from the building up of material on a depositional surface such as the bottom of a lake or sea. The concept that the younger material rests on top of the older seems so obvious as to be almost trivial, and geology students sometimes feel intellectually insulted when they are asked to remember that in 1669 a Dane named Niels Steensen published a treatise that told the world this most obvious fact, now dignified as the Law of Superposition. The factors that influence thought were not, however, at work in Steensen's time. For one thing, even though Gutenberg had invented movable type some two hundred years earlier, printing was not big business, and there was not the mass of paper that we have today to provide Steensen with a desktop model of sedimentary rock deposition. Furthermore, at that time people generally believed that the world was formed in a series of a few cataclysmic events, not built up gradually by the accumulation of anything.

Steensen was a biologist and a physician, a naturalist if you will, who attracted the attention of the rulers of that most Renaissance of all cities of time, the Florence of the Medici. In 1666 Ferdinand II de Medici invited Steensen to be his guest and provided him a house and all the necessities of life a scientific genius in the care of a rich and powerful patron might require. Niels was nothing if not adaptable, having done research in Holland and France on his southward migration from Denmark, and he promptly changed his name to Nicholas Steno and became a Catholic. The latinization of his name and his affiliation with the local church no doubt helped somewhat to protect him from attacks by the church hierarchy as he continued to pursue his studies of anatomy, a line of work that was considered heretical by the powers of the time, as many other scholars had found out to their misfortune. Steno, though, was Ferdinand's personal physician and the tutor of his sons as well as a frequent traveling companion, so his protection from the ire of the church was secure.

Steno began his geologic studies by comparing fossils of

shark teeth he found in the rocks of Tuscany with those of modern sharks. He came to the conclusion that the teeth from the rocks were not simply oddities of the inorganic world, as was thought, but were indeed remains of extinct animals, the ancestors of today's sharks. He also studied the rocks of the area, putting his observations and ideas together in a small book that would one day be praised as a classic in geology and paleontology. This book was the last of Steno's scientific work for he then became a priest, later a bishop, and served in a number of high church positions in Germany. He was finally transferred to the tiny duchy of Mecklenburgh and in the end died in poverty. The result of the rapid transfer of his interests and activities to the church meant that his scientific ideas did not get the widespread attention that they might have until many years after his death.

One can assume that Steno, even in his most difficult last days, was able to keep the organization of his papers straight by applying the Law of Superposition. It is tempting to speculate that this brilliant man might have been demoted in the church pecking order when a superior construed his desk piled high with papers and books as a sign of an unstructured mind, thinking wrongfully that an executive should have an uncluttered desktop.

Thus it is with over three hundred years of precedence that geologists use a stratigraphic system of filing. Those who are good at it can reach into a pile and find the paper they're looking for within a page or two. One of my professors at Berkeley—the volcanologist Howel Williams—used his incredible memory to set up what could be called a quantified system of stratigraphic filing in which he simply numbered copies of scientific papers as he acquired them, in serial order, and then remembered the number for each, at least within a digit or two. While I lack the almost total recall of Williams, my students were always amazed at how easily I could locate an old exam or a letter out of the piles on my desk and bookshelves. It is simply the geologists's mind at work, following Steno's lead.

Whether it is a pile of rocks or a pile of papers, as long as the pile is not disturbed, Steno's powerful though simple law serves to establish a relative time scale among the layers in that pile. But most scholars of Steno's time were content to force what they observed in the rocks into the Noachian concept of a Great Flood so as not to disagree with the powerful church leaders of the times. Leonardo da Vinci had proposed a modern view of fossils in his notebooks of the mid 1400s, but his works and Steno's remained in obscurity until well into the days of modern science.

Tuscany must have been a fertile place for geologic observations, for it was there in Steno's old region of discovery that an inspector of mines named Giovanni Arduino studied the various rocks exposed in the hillsides and mine workings that were his turf. Arduino came to the conclusion that they could be divided into several broad types that had distinctive characteristics and were of different ages. In 1760, almost a century after Steno's booklet of 1669, Arduino proposed that there were four main types of rocks in Tuscany:

Primary: the oldest rocks, which contain mineral deposits and are devoid of fossils;

Secondary: younger rocks that are well stratified and consolidated and contain fossils but no ore deposits;

Tertiary: still younger stratified rocks that are unconsolidated and contain numerous marine fossils; and

Volcanic: rocks such as ashes and lavas that generally are interlayered with the Tertiary rocks.

Arduino also proposed a category called *Alluvial* for the coarse and bouldery deposits that were obviously laid down by modern or near-modern rivers or by shoreline action. His categories have stood the test of time, although they are not used today in exactly the same sense as he used them. Primary rocks are the crystalline rocks, the granites, gneisses, and schists (generally coarse-grained rocks made of feldspars, quartz, and micas or other dark minerals) of the mountains; these are the so-called basement rocks of later generations. Secondary rocks include the layered, sedimentary rocks of the mountains, rocks of the Mesozoic age. Arduino's use of the term *Tertiary* has survived more or less intact to describe the bulk of the post-Mesozoic rocks (rocks of the last sixty-five million years) that are found in the world, and his Alluvial rocks would be called Quaternary or Recent today, though the term *alluvium* is still used to describe unconsolidated rocks of young age.

Various scientists in Europe followed Arduino and began to divide rocks in their local regions into categories of different ages, though there was no explicit indication that the rocks of those areas could be compared with those of Tuscany. Johann Gottlob Lehman studied rocks in Saxony in Germany and recognized that the earth was not a disordered jumble but made of layers or strata. In Berlin in 1756 he published a book, *Versuch einer Gesichte von Flötzgebürgen*, that contained an almost complete cross-sectional view of rocks that we now call Permian in age, which he studied near the Harz Mountains. He also presented a classification similar to Arduino's (not to mention some pretty wild ideas about the origin of mountains). Lehman's book was also published in a French translation in 1759. A German contemporary of Lehman, Johann Christian Füchsel, studied a succession of rocks in Thuringia, separating them into units on a detailed scale not unlike our present formations, which are extensive units of uniform character. Neither Lehman, Füchsel, nor any other workers of the time hade any attempt to extend their classifications beyond their local areas of study.

That step was left to a fellow named Abraham Gottlob Werner, who was the son of a mine owner and one of a long line of people in the minerals industry. In 1775 he joined the faculty of a then-obscure mining academy in Freiberg at the age of twenty-six after graduating from the University of Leipzig. Historians tell us that Werner was an unusually skilled orator, presenting entertaining lectures to large groups of admiring students from all over Europe who believed every word he said with uncritical awe (the sort of person who might win a good-teaching award today). Unfortunately, though Werner did apparently go into the field, he seldom went very far from home. His field work led him to the conclusion that all the rocks around Freiberg and in Saxony were formed by precipitation from the waters of a universal flood, thus following Lehman's theory, and he mistakenly included in this category many rocks that are now known to have been formed by the crystallization of molten silicate liquids or the metamorphic reconstitution of preexisting rocks at high temperatures and pressures.

Werner extrapolated his ideas about rock systems beyond Saxony and published his "Short Classification" in 1787 in which he stated that all rocks, *everywhere*, were deposited from water and that they were Primitive rocks, Transitional rocks (a group added after the initial publication of the work), and a third group called Stratified rocks, with an alluvial class to include things like peat, ashes, clays, and the like. He had found what we now call Paleozoic rocks (his Transitional rocks) to fit in between Arduino's primary and secondary units. He also later added a volcanic group to include volcanic rocks of obviously recent origin. Werner hypothesized that the older rocks that were contorted and often found standing on edge rather than in more or less horizontal layers had crystallized out of solution along the steep slopes of an ancient earth core. For later units, Werner allowed that there was some participation of sedimentation of rock particles in the deposition of these younger types. Werner insisted that these classes were worldwide in extent and application and had enough people believing him to hold up progress in geology for a few decades. Because of the emphasis on the importance of worldwide seas, the followers of Werner came to be known as Neptunists.

Werner called himself a *geognosist*, a term coined by Füchsel, by which he meant a scientist of broad interests, one concerned with all aspects of the physical earth. He also believed that nature was orderly as a matter of principle, not unlike Lord Kelvin. Werner's beliefs simply led him to generalize too broadly. We now view his beliefs that Primitive rocks were precipitated from solution in a primeval sea and that the great sea subsequently vanished into space as utter nonsense, but we should remember that Werner had a life-long familiarity with veins in mines, which were filled with layers of crystalline minerals that had obviously been precipitated from solution. Also, he did conceive the idea of the correlation of rock units over vast areas, and as A. Hallam of the University of Birmingham notes, it was the *idea* of world-wide correlation that excited scientists. Werner was altogether wrong about most of his correlations and about the origin of many rock types, but his grand conception of universal units inspired his followers to study rocks all over Europe to support their mentor's erroneous views. One by one they left Werner's camp as they found evidence contrary to his great scheme, but it was his inspiration that had sent them to the field, and in the end the work of geology had been advanced and was rolling full steam ahead.

Werner did help geology by classifying rocks, and his classification of minerals provided order to a field that was a chaos of terms. The value of his concept that rocks might be equivalent over vast areas and the inspiration he provided to students cannot be underestimated. But he did not help the *actual practice* of the correlation of rocks—the demonstration that a rock in one place is part of the same layer as another rock in another place. Neither did he demonstrate that a rock in one place is the *time* equivalent of one in another place. Werner's big mistake regarding correlation—his conclusion that the same rock types in two regions *must* be the same age—was one of the fundamental errors of geology, an error that persists to this day in the thinking of many geologists. If we return to the model of the geologist's desk, you should see the principles clearly.

Let's suppose that a number of geologists share a large office at a company or a university, and that each of them has a pile of papers with a pink sheet in it. By the simplest Wernerian principles we would say that each sheet was the time equivalent of, and could be correlated with, every other sheet. If we looked at the sheets and found them to have different material printed on them, then we might refute the correlation. Suppose we looked at them and found them to be copies of the same office memo. Can we correlate then? Werner would say yes but I say no. The pink sheets may be duplicates that were printed at the same time, but suppose that they were taken from a file that was circulated. Each pile would have the same item, yet each one would have been added at the time the geologist took his copy from the file and kept it. The identical items would then be added or deposited over an interval of time, with no two of them time equivalents. These items can be correlated in the same sense that they are all parts of the same unit, but each one was deposited at a different time. You must be careful not to confuse the *time of deposition* with the *physical identity* of units in a set of layered piles of either papers or rock.

Another sheet might contain information to which the geologists might refer now and then so that it would be recycled to the top of the pile at intervals which would differ according to the needs of each geologist. Its original date of formation would be of interest, but where it was in the pile would depend on when it was last recycled, much as a rock can be relocated by such movements in the crust as faulting or folding, as Steno recognized.

There might be some cases for which time correlation within the piles of papers, based on physical characteristics, could work. Suppose, for example, that one of the windows was left open on a windy night and some dust blew in so that there would be a film of dust on a given sheet in each stack. The dust layer would be a time plane connecting all the piles of paper. Such events would leave their mark, but what if dust had settled on the piles at two or three different times or some of the geologists had blown the dust off and some hadn't? Then you could not be sure.

The geologic equivalent of dust blowing in the window is volcanic ash falling from the atmosphere. The deposits left by this event serve as markers that can define time planes within sedimentary rocks. The ash forms a rock called tuff that is commonly found as a thin layer in exposed rock sections over vast areas, representing what may be imagined as an extensive tabular body of rock. Such a layer can occur within accumulations of diverse types, and, if it can be positively identified, is very valuable as a time correlation marker. It may be present in marine rocks, in rocks deposited in lakes, and in rocks laid down on land, thereby indicating that the three different accumulations were laid down at the same time. The ash is not literally deposited everywhere simultaneously because the ash cloud may take hours or days to travel from one place to the next, but in a geological context that is all but instantaneous. You should also recognize that the marker layer may not be present in every exposure of the rocks, even at the right time level, because some of the volcanic ash might have been removed by natural processes of erosion before it was protected by burial, just as with the dust layers on the geologists' desks.

Because time planes like ash-derived tuffs are so useful, geologists seek them out and use them avidly when discovered. The trouble is that many tuffs and their derivatives are identical in appearance to other, quite unrelated tuffs. A tuff unit was found in the Pleistocene rocks of the western plains and named the Pearlette Ash. Geologists thought that they had found a correlation tool that required no work at all. Just find the Pearlette and you know where you are in the stratigraphic pile—no fuss, no muss. There was only one problem with the Pearlette. There was more than one ash layer being called Pearlette, and nobody could tell one from the other.

There are tuff layers in some rock accumulations that do serve as time planes over limited areas, but they all suffer from identity problems. Unless they can be traced from outcrop to outcrop on a nearly continuous basis or contain highly unusual particles of some kind, they are suspect.

Some earth scientists and physicists proposed a few years ago that there is a worldwide time marker at the end of the Mesozoic (about sixty-five million years ago) that is the result of the collision of the earth with another cosmic body. This marker, an unusual concentration of the rare element iridium, has been found at what is purported to be the same stratigraphic level in a number of localities throughout the world, but at present the geologic community is polarized on the question of the origin and significance of the iridium. Recent evidence suggests that the iridium could have come from terrestrial volcanic sources. Also, there is no certainty that there are not other, undiscovered layers in the rocks of the world with comparable concentrations of iridium. U.S. Geological Survey geologists have recently reported the presence of quartz grain in the boundary claystone that are interpreted to show characteristics caused by extreme shock as would arise from impact. This evidence is not conclusive because such quartz might form in other ways, but it may prove to be important. It is too soon to call this one.

Apart from events like ashfalls and the possible impact of extraterrestrial bodies, there are really no natural circumstances that give rise to time markers of a truly global scale that could be preserved in sediments of diverse types. Many rocks do preserve a record of the orientation of the magnetic field in which they formed because magnetic mineral grains retain a magnetic memory much as do the particles of magnetic material on a recording tape or computer disk. This so-called remanent magnetism can be used as a time marker of sorts because the magnetic polarity of the earth's magnetic field reverses from time to time. These reversals can be located in a column of sedimentary rocks and can be correlated with one another from place to place. Unfortunately, it is a bit like the Pearlette situation in that one magnetic reversal cannot be distinguished from another in most instances.

Rock types generally reflect their conditions of formation rather than the time of formation, which is the principal monkey wrench in Werner's grand scheme. A widespread example of a rock type that reflects conditions of deposition is rock formed from beach sands. Beaches are places where a lot of winnowing and mechanical abrasion take place. As a result a separation occurs in which finer materials are washed away and the coarser sand is left on the beach. In the case of most beaches around continental masses, the sediment supplied to the beach has been transported long distances by rivers and consists mostly of the chemically and mechanically stable mineral quartz. In other words, if you've seen one beach sand, you've seen them all. The beach sands around islands and the like are exceptions because they are commonly made of locally derived grains such as carbonate or volcanic materials.

Although not particularly common, beach sands are easily recognizable and a geologist could be tempted—perhaps by the ghost of Abraham Werner or, more likely, by the same wish for simplicity that led Werner astray—to use a beach sand layer as a time-correlation indicator. I suppose it fair to say that a beach sand layer does serve a purpose in the geologic record in that one can be reasonably sure that it crosses time as it is traced from place to place, at least in most directions. To understand this statement you must first realize that a beach is a very narrow strip of sand that lies along a coast and has a width that is bounded by the high wave line on land and by the place away from shore where the winnowing action of waves ceases to be effective because of water depth. A beach sand body may extend for hundreds of miles along a coast but is generally only a few hundreds of feet or at most a few miles in width.

The only way a layer of beach sand could be deposited over a large area would be if the relative height of the landmass and the level of the sea changed with time. Suppose that the level of the sea rose slowly (or the landmass sank). The beach would then migrate slowly inland as the water buried the previous beach sands. Under suitable conditions the old beach sands could drop below the zone of wave action and be preserved, generally buried by finer-grained sediments as they became covered by deeper and deeper water. As the beach moved landward the original line of sand would become a ribbon, then a sheet.

Think about the relative time of deposition of the different parts of this seemingly homogeneous layer of sand. Let's imagine a beach along a more or less straight shoreline of an inhabited region that is gradually being inundated by the encroaching sea. The people who live in this land are known as the Calendaphiles because they believe that calendars are sacred and make them of stone, with a slab for each year.

Their priests have told them that the reason the sea is flooding the land is that the gods are angry because some disbelievers have been saving old calendars to use when the dates and days coincide in future years instead of making new ones for each year. The priests say that the only way to prove their faith is to cast their old calendars into the sea at the end of every year instead of squirreling them away. As a result the beach sands contain a convenient marker, and a scientist studying the sand layer at a later time would find that the linear parts parallel to the shoreline were of the same age and that at right angles to those lines the sand layer was older in a seaward direction and younger in a landward direction. Unfortunately, geologists don't have any time references quite as handy as the calendars of the Calendaphiles, but the principle is the same for any sort of deposit, beach or otherwise, that forms by transgression of the sea over the land. (If the sea level were retreating, the beach would migrate seaward, and the time relations would be the reverse of the case stated above.)

In this beach sand example we see an example of old Werner's ideas at their worst. Even a continuous layer of rock is not the same age at different places. Put another way, the beach sand layer is a rock unit, but it is not a time unit; the layer represents an environment of deposition, not a time of deposition. This obvious fact is not at all well understood by many practicing geologists even today, many of whom were taught stratigraphy of the type called "layer cake" in which correlation is accomplished strictly according to the Wernerian principles of correlating rock types or rock successions with one another on the basis of physical characteristics.

There have been times during geologic history when conditions were very similar over large areas, and Wernerian or "layer cake" correlations would work in a loose way. The special sort of iron ore deposited during Precambrian time can be roughly correlated from place to place because it represents conditions of formation that have not recurred. In the Rocky Mountains the Morrison Formation, the burial place of most of the dinosaurs found in museums today, looks identical in parts of Colorado, Utah, and Wyoming that are hundreds of miles apart and was probably deposited at about the same time everywhere over a land surface of low relief. There are many more examples of the situation, but they are the exceptions.

Werner's concept of rocks layered one upon the other in ponderous orderliness is not supported by the evidence, and we can see that his theory was a combination of limited observations and a lot of wishful thinking. No matter how hard we try, we find no support for water appearing from nowhere, depositing rocks, then vanishing and leaving sediments that are time equivalents everywhere. We instead see even today that carbonates are being deposited in the Bahamas, diverse sediments and lavas in the ocean deeps, beach sands in North Carolina, tuffs in the Pacific margin, and glacial deposits off Greenland. The great majority of the geologic record is likewise one of local strata, most of which transgress time in the area they cover.

By way of a historical aside, Werner's misinterpretation of the origin of many rock types put him at odds with Hutton's followers who became known as the Plutonists (after the Greek god of the underworld) because of their belief in the deep, igneous origin of crystalline rocks. The Neptunists lost the battle over the origin of rock types, but the Plutonist contribution to stratigraphy was virtually nonexistent—they had little or no stratigraphy, only a broad theory of earth history.

The momentum provided by the work done by Werner's disciples together with the widespread use of the new concept of using fossils for correlation (following the lead of William Smith and others in the early nineteenth century) led to the rapid development of a worldwide stratigraphic system that did work, or at least *appeared* to work. Correlation based on the identity of fossil assemblages contained in a pair of rock units was the basis for constructing the relationships among the rocks of Europe and England. If two apparently similar rock units contained the same fossils, was it fair to say that they were part of the same unit? The answer is: maybe. William Smith used correlation of strata by fossils over *short* distances and was therefore probably correct in his identification of the same layer in different places, but Smith did not understand that a single stratum can transgress time, as did the beach sand of the Calendaphiles.

There are three kinds of correlation that can be done using fossils: the correlation of both time and environment or the correlation of either factor separately. Let's return to the example of the geologists' office once again for clarification.

The easiest event to visualize is one which is recorded simultaneously everywhere in time alone. This is the case when the custodian checks the office during the lunch break on a summer day and finds that some windows have been opened to let in a breeze and that a lot of flies from a nearby hog farm have flown in and are buzzing around the room. He quickly closes the window, gets an insecticide can, and sprays the office. In a few minutes most of the piles of papers on the desks have one or more dead flies on top of them— fossils if you like—killed and preserved in an instant of time. The different piles of papers represent many distinct environments within the broad environment of an office in summer fly season.

In a slightly but significantly different situation, let's suppose that the custodian noticed the flies on a Monday morning when all of the geologists were working at their desks. He would realize that he couldn't spray the place with the people there, so he would quietly hang an insecticide strip at one end of the long room. Over the course of hours, days, weeks during the breezeless summer the vapors would spread down the room, and the flies would be deposited on the piles of papers at successively younger times as the wave of insecticide swept onward on its deadly mission.

Whereas the first layer of flies represents time, the second one represents only the environment of an office in summer fly season. The stratigraphic position of the second fly layer, considered over the room, transgressed time as it formed. A scientist arriving later to examine the distribution of dead (fossil) flies would be hard pressed to distinguish between the two cases without using evidence other than the flies alone, such as, perhaps, dates on the papers in the piles.

The custodian in the geologists' office always tries to do a good job of controlling flies in the whole building, not only in their office. To that end he places fly traps around the building in various places, replacing them when they become ineffective because of dust or drying out. The flies trapped represent the environment of fly traps and signify nothing at all about time both because the traps are hung at different times and because the length of the time interval represented by each trap is different from the others.

To return to geology, let's look at some examples that might be found in rocks instead of on desks.

The flies that were all killed at a single time may be compared with any animal that is highly mobile and thus can be expected to be preserved at many places in many rock types. Migratory birds would, theoretically, be an ideal fossil for time correlation because they would have a nearly equal likelihood of falling to earth almost anywhere, although in practice they are seldom preserved as fossils. In the oceans mobile species are useful correlation tools, and the swimming molluscs called ammonites have been used for worldwide correlation, especially in rocks of the Jurassic and Cretaceous periods, where they are abundant as fossils.

The flies killed over a span of time are similar to creatures that can live only under a narrow range of conditions (such as temperature or salinity). As conditions at any point change, the organisms must either move or die off. For example, as the beach sand layer of the Calendaphiles built landward it would cover remains of terrestrial organisms such as plants, mammals, land snails, and the like that could not survive under water. Also, organisms adapted to life in the active beach zone would migrate inward with the beach so that the same fossils would be found at different places in the beach sand layer but would not represent the same time. On a large scale, because microscopic organisms in the oceans are distributed according to the temperature of the water, the distribution of their fossils in marine rocks transgresses time, showing instead how ancient conditions slowly changed. With slow-moving animals such as mammals, their distribution as fossils commonly represents the speed with which a species spread to new territory over millions of years (much as if the flies entered through a small opening at one end of a long room and were collected on fly paper as they spread away from their source).

The case of the flies caught in the fly traps, which represents a specific environment, is similar to some fossil assemblages studied by the French paleontologist A. D. d'Orbingy. D'Orbingy found occurrences of corals and other reef building organisms in a number of places and considered them to be time equivalent. In fact the equivalence was not time but water depth, temperature, food availability, and other factors.

William Smith had no concept of how complex correlation with fossils really was, nor did anyone else for many years to follow. Fossils are invaluable tools for correlation, but you have to be sure what is being correlated. The identification of fossils, together with a detailed knowledge of the physical characteristics of the rock that encloses them and radiometric dating, can give today's geologist a full set of tools to work with to determine both age and environmental factors, providing he or she is willing to work with *all* the data.

I suppose that there will always be an ethnic flavor to any discussion of Werner, and I found that a German-educated colleague of mine objected to an earlier version of this essay in which I really took Werner to task. Likewise it is easy to criticize the flamboyant Frenchman d'Orbingy for his naive correlations of reef assemblages. Such is nationalism. As for me, I'll be an objective scientist and side with my Scottish and English ancestors, except of course for the ones who were Neptunists.

Chapter 7

Better Paleobiology through Chemistry

MUCH OF THE EVIDENCE USED TO SUPPORT EVOLUTION ARISES from observations of modern living things, but the principal source must by necessity be the fossil record, for it is only there that we can examine parts of plants and animals that once lived on the earth and gave way to successors. The breeding of animals and plants and studies of genetics and inheritance even down to the molecular level only tell us that evolution *can* happen. It is the fossil record that persuades us that it *did* happen.

The fossil record is, alas, incomplete. First, generally only the hard parts of organisms are preserved: the shells of molluscs (clams, snails, etc.), the bones of vertebrates, the wood and seeds of plants. It is highly unusual for an impression of soft tissue to be preserved, and even though the texture of a

For more information, see references 14, 23, 31–34, 42, 51, 54, 55, 57–63, 69, 86, and 96 in *Further Reading.*

dinosaur's skin has on occasion been found preserved as an imprint in sedimentary rocks, we know nothing of the nature of its skin cells or what its pancreas might have looked like, or even if it had such an organ. Second, there is not a record in the rocks of *every* species that existed at *every* instant in past times. There is instead a rather spotty record of some common organisms that had hard parts that were buried soon after death in an environment that preserved them as fossils. We know from studies of modern environments that the preservation of organic remains is highly unusual. Even the rocks that originally enclosed fossils commonly become melted, folded, and crushed; crystallized into new rocks; or reexposed at the surface and destroyed.

The idea that we should find every intermediate step in the evolution of one form to another is naive and old-fashioned, and Darwin's insistence on gradual change made his defense of organic evolution more difficult than it would have been had he allowed for rapid change instead of modeling evolution after Lyell's concept of gradualism. Both because there are gaps in the sedimentary record and because organisms seem to change during brief bursts following times of little change (stasis) through a process called punctuated equilibria by paleontologists Stephen Jay Gould of Harvard University and Niles Eldridge of the American Museum of Natural History, it seems reasonable to conclude that evolution proceeds by abrupt changes. Anyone who has bred domestic animals knows that a whole flock doesn't change through selective breeding without culling, any more than an entire species population changes into another. What happens is that one individual animal or a small band becomes the basis of a new breed in the barnyard or a new species in nature, and such changes are rapid.

From this incomplete record which contains gaps both because of lack of preservation and because of rapid evolutionary bursts, paleontologists have constructed numerous family trees, mostly with dead ends, and have conceived a probable sequence of organisms that lived during the last half billion years. Most of the evidence comes from preserved hard parts, though animal tracks have been found that provide information about locomotion and social habits. We have reached some conclusions about the dietary habits of extinct animals based on the structure of teeth and jaws. In fact, many extinct animals are known only from their tracks, a unique skeleton, or even a single tooth.

In spite of (or perhaps because of) this imperfect record, there is general agreement among specialists about the lineage of fossils and modern species. New fossils are still being discovered, but the evidence is finite. It would be nice if there were means other than physical examination of testing accepted beliefs about the lineages derived from fossil morphology.

It was just such a desire that led paleontologists to dabble with chemical means of studying fossils. It was found early on that fossils were composed mainly of carbonates, phosphates, or silica, but such broad compositional generalities did little to foster better understanding of them. From the years following World War II to this day, there has been an explosive development of new instrumental methods of chemical analysis that not only has led to a greater sensitivity and accuracy in the analyses but has also made the analytical procedures much simpler and therefore cheaper and more rapid.

For paleontology this revolution resulted first in an embarrassment of uncritical analyses of lots of fossils to detect whatever chemical elements happened to be within the capabilities of the instrument available to the investigator. The new toys were exploited to gather data that nobody knew what to do with. Lots of very silly and stupid scientific papers were published.

Still, any new field has growing pains, and the workers continued to gather data and then seek explanations for what they stumbled upon. One important question that these "pioneers" avoided was how to distinguish between chemical elements that had become incorporated during the life of the animal and those that were added later during the process of fossilization. Insofar as telling us anything about the living animal it is only the antemorten, or predeath, chemical composition that is potentially useful. Happily, there were brandnew analytical instruments just coming on the market that were capable of solving this question. The electron microprobe analyzer and the scanning electron microscope with analytical capability enable the operator to make chemical analyses of tiny volumes of a sample. Regions of a fossil only a few micrometers across could be analyzed (a micrometer equals one-millionth of a meter, or 0.000004 inches). Elements that were parts of pore or crack fillings in the fossils could be distinguished and were thereby shown to be postmortem additions to the fossil and not a relict of the living animal. Heinrich Toots of the C. W. Post College Geology and Geography Department was a graduate student at the University of Wyoming in the early 1960s and had made survey analyses of a large number of vertebrate fossils as a first step in fossil chemistry studies. Toots and I selected a suite of key specimens and I obtained research funds to travel to the Pomona College Geology Department to use their electron microprobe analyzer with the bone and tooth samples we had selected. I was able to demonstrate that silicon, iron, manganese, and a number of less common elements had been added as postmortem fillings of voids in the fossils.

A number of other elements, notably fluorine, sodium, and strontium, were confined to the solid part of the fossil bones and teeth, and we strongly suspected that the latter two had been incorporated into the bone during the life of the animal. (I'll have more to say about strontium in Chapter 10.) Bones and teeth are composed of a major mineral component and a minor cellular, organic component. The mineral component is a substance called *apatite* by mineralogists (see Chapter 1). It is composed of arrangements of oxygens surrounding phosphorus atoms called phosphate groups. The phosphate groups are in turn associated with calcium and with the pairing of hydrogen and oxygen, called hydroxyl, the whole structure forming a stable crystalline material. Practically no real-life apatite is composed solely of phosphate, calcium, and hydroxyl, however; it also contains a host of other elements. These other elements can substitute for the major elements in the apatite crystal structure because they fit comfortably into the same space as a calcium, hydroxyl, or phosphorus and have suitable chemical properties that allow them to be chemically attached, or *bonded*, to the other atoms in the structure. Thus a chemically pure apatite consisting of calcium, hydroxyl, and phosphate exists only in the mind of a mineralolgist.

In apatite both strontium and sodium have the right combination of size and chemical properties to allow them to substitute comparatively freely for calcium. Strontium forms a very strong chemical bond with the apatite structure and consequently is not readily removed during the postmortem processes that affect a fossil bone or tooth.

I had found considerable sodium in the solid part of some modern bones and teeth, so Toots and I concluded that the sodium in fossils was introduced during the life of the animals. Furthermore, since the chemical bond of sodium with apatite is relatively weak and it is highly soluble in water, Toots and I suspected that sodium would be readily removed by chemical leaching during the fossilization process.

The other common substitution in bone apatite is the replacement of hydroxyl by fluorine. The fluoride ion is about the same size as the O-H pair that constitutes hydroxyl and also forms a very strong chemical bond with the apatite. Consequently fluorine has a very strong chemical affinity for the apatite structure and can occupy all the hydroxyl sites. The result is a mineral called fluorapatite. In fact, the bond of the fluoride ion with the apatite is much stronger than that of hydroxyl for which it proxies. This is the rationale behind the use of fluoride toothpaste or fluoride-containing drinking water; the strong bonding of the fluoride ions makes the apatite crystals in tooth enamel and dentin less susceptible to attack by decay processes.

Fluorine levels in modern bone are nearly nonexistent compared with those found in fossil bone, so Toots and I assumed that virtually all of the fluorine in the fossil bone apatites had been added postmortem. In any case it had been known for some time that fluorine levels increased with the age of fossilized bone, and anthropologists had attempted to use fluorine levels in fossil bone to date the specimens. The method proved to be highly inaccurate for absolute age measurement but was at least accurate enough to expose the Piltdown Man hoax when it was used by Kenneth Oakley in 1950 to show that the fluorine content of the two parts of the supposed fossil were so vastly different that they could not possibly have been parts of the same individual.

Some time after I had done the electron microprobe work on Toots's samples he and I analyzed fossil samples for both sodium and fluorine, elements we had previously been unable to measure because of the limitations of the X-ray spectrometric methods we had been using. As mentioned above, we suspected that circulating water would readily leach sodium from the fossils while at the same time introducing fluorine. To quantify our educated guesses we selected some suites of bone and tooth samples that were all from a single skeleton, and hence identical in terms of age and conditions of burial, though they were quite different from one another in terms of their porosity and permeability. With age and conditions of burial fixed, any differences among the samples could be attributed to the physical properties of porosity and permeability. We were interested in these properties because high porosity favors chemical reactions such as the leaching of sodium and the addition of fluorine, and high permeability has a similar effect in that it speeds the passage of groundwater through the buried fossils.

We found that the most porous and permeable bones contained the least sodium and the most fluorine and that the very solid and impermeable tooth enamel contained the most sodium and the least fluorine. These results were exactly what we had predicted.

The results put a final nail in the coffin of the fluorine dating method, for Toots and I found more variation in fluorine content within a single skeleton than would normally be found between fossils of different ages from different localities. The fluorine method will doubtless still receive some use by anthropologists just because it is so simple that the temptation to employ it is difficult to resist. One member of the audience who heard me present our study that totally discredited the method, nevertheless asked me pleadingly if it still might be used in some circumstances. I told him that I wouldn't, but I'm guessing that he will. I should point out that Oakley was able to expose the Piltdown man fraud because one part of the supposed fossil was shown to be modern rather than a fossil because it contained essentially no fluorine at all.

Toots and I believed that studying the sodium content of fossils still had great potential, for if we could somehow get good estimates of the sodium levels in the fossils prior to any leaching we would have an insight into the physiology of the animals. Sodium is the major extracellular electrolyte (an electrolyte is a soluble chemical entity with an electric charge) in vertebrates, and its concentration in bone should be proportional to its concentration in body fluids. A knowledge of its levels in extinct animals would tell us something about their metabolism, information that cannot be obtained in any other way.

Toots and I needed two things to allow us to estimate life levels of sodium from fossils. First, we needed a skeletal element that had suffered the least possible change after burial, and second, we needed some way to correct data from those samples to arrive at a life sodium value.

Tooth enamel comes closer to the first requirement than any other part of the skeleton for a couple of reasons. First, enamel is different from other skeletal tissues in that the tiny apatite crystals that make it up are about a hundred times larger than the crystals in bone or tooth dentin and larger crystals mean lower chemical reactivity. Second, enamel is almost devoid of organic material and consequently is far less porous and permeable than bone or dentin. We had also found out empirically from analytical data that enamel was the least affected by postmortem change of any of the tissue types in vertebrate fossils.

We could reasonably suspect that enamel might be chemically different from other bone because of its different physical properties, but both we and others had shown in previous studies that strontium, an element that is incorporated into the skeleton during the life of the animal, is present in the same amounts in enamel as in other bone tissues. Thus Toots and I could reasonably assume that the sodium levels in enamel would reflect skeletal levels and hence body levels.

Enamel is not completely resistant to postburial change in sodium content, so we still had to correct our data to arrive at life levels. What we needed was an element that was added to the fossils at a rate that was in inverse proportion to the rate at which sodium was leached out. Such an element would have to be almost universally present in groundwater and relatively unaffected in its chemical properties by minor variations in water chemistry. Fluorine seemed to meet the qualifications.

We tested our theory by plotting the fluorine content of some enamel samples from a single species taken from a single locality against the sodium content of the same samples. In doing so we were holding the variable of groundwater composition constant since it was assumed to be the same in all parts of one limited locality. We held the life sodium level constant by selecting animals of the same species because we assumed that all the individuals of the species would have had a similar diet and metabolic makeup and hence would have comparable life levels of sodium in their enamel. Thus we were testing whether there was any correlation between sodium leaching and fluorine addition. To our delight the points defined a straight line with very little scatter of the data, which indicated that sodium leaching was indeed proportional to fluorine addition for that suite of samples. Since life levels of fluorine are almost negligible (about 0.002 percent), we could estimate life sodium levels by projecting the line back to zero fluorine. We did not actually use a graphical method but, using standard statistical methods, computed an algebraic expression based on the fluorine and sodium data and then used the expression to estimate the sodium content of a sample with zero percent fluorine, the estimated life sodium level of the bone for the species. It is well to remember, though, that just because there is a correlation between two sets of data, it does not necessarily mean that there is any relationship between them in the physical world. Additional tests were necessary.

I got my research assistants and chemist busy preparing and analyzing more enamel samples so Toots and I could test our theory further. As we had hoped, the new data defined straight lines just as had our initial trial, with the data from each species or other taxonomic group of animals giving a different line. We plotted all the lines on a single sheet and discovered that they intersected at a point that had the coordinates 3.8 percent fluorine and 0.1 percent sodium. This confirmed that the relationship we observed could be attributed to a real physical process. The fluorine value was in agreement with the fluorine content of a theoretical apatite in which all possible fluorine sites in the structure (the hydroxyl sites) are occupied by fluorine. Thus it represented the maximum possible fluoridation of the bone apatite by postmorten change. The sodium value was in agreement with the lowest values we had ever obtained, those being from highly porous and permeable bone samples that also had a physical appearance suggesting extensive alteration after burial. Thus 0.1 percent sodium appeared to represent the maximum possible effect of leaching. We could therefore say with confidence that the lines defined by our fluorine and sodium data showed the pathway by which the composition of fossils changed, with increasing alteration due to the processes of sodium leaching and fluoridation, processes that tended toward a terminal composition of 3.8 percent fluorine and 0.1 percent sodium.

Why wouldn't all of the sodium be leached rather than just most of it? There are numerous ways to look at this question, but they all come down to the energy states of sodium ions and apatite crystals. Consider the simple case of table salt dissolving in water. It will dissolve until the brine becomes saturated, which is just a circular way of saying that no more solid salt will dissolve in it. A more instructive way of viewing the process is to regard it as two competing reactions, one of solid salt going into solution and another of dissolved salt forming solid, crystalline salt. As long as the *rate* of solution exceeds that of crystallization the net result will be solution. When the rates are the same, there is no net solution or crystallization; that is the saturation state. In other words, in the saturation state the total energy of a sodium ion is the same in solution as in a crystal of salt, so any given ion would not tend to change its state from solution to solid. All chemical reactions proceed in the direction of lower total energy, and if there is no difference between two energy states, there will be no net reaction.

Applying the same principles to the apatite in the fossil we can predict that sodium will tend to go into solution from the apatite—in other words, it will be leached—as long as the total energy of sodium ions in the apatite is higher than in the groundwater. All groundwater contains some sodium, so we would expect to find that the groundwater will be an effective solvent for bone sodium until the amount left in the bone becomes so small that the energy of the sodium ions in the fossil is equal to their energy in the dilute sodium solution that is the groundwater. That level is approximately 0.1 percent sodium in the bone apatite.

The theoretical end point of maximum fluoridation and sodium leaching not only confirms that the sodium-fluorine correlation is the result of postmortem alteration but is also useful regarding data that don't yield such clear-cut results. For example, we found as expected that very thick tooth enamel was resistant to alteration, and as a result the plotted sodium and fluorine values did not spread out sufficiently to define a very believable straight line. In other words, the scatter of points looked more like a shotgun pellet pattern than a linear arrangement. With these we simply put a step into our computer program for line fitting that forced the line to pass through the 3.8 percent fluorine and 0.1 percent sodium point, thereby giving a line that could be projected back to zero percent fluorine in order to estimate life levels of sodium.

Fortunately most of the data sets required no such mathematical trickery to define straight lines. One especially intriguing plot was of data from the horse *Equus* (the same genus as the modern domestic horse) that lived in North America during the Pleistocene epoch. Toots and I first plotted data from two different age levels at a Colorado locality and got a well-defined straight line. We obtained some more *Equus* teeth from another locality in Nebraska and found to our suprise that they plotted on the same line even though the localities were hundreds of miles apart and not even exactly the same age. Next we analyzed some *Equus* enamel from a Florida occurrence of Pleistocene age and were both astonished and delighted to find that they too plotted on the same line. We knew then that the sodium content of the enamel is related to the kind of animal, not to its diet, environment, or other factors. It told us something about the animal's inner workings.

We then gathered data from a great variety of different genera of horses and found that each genus defined a separate line. We also found that estimated life sodium levels were related to geologic age and hence to the evolutionary stages of the horse. In a systematic way, the higher a horse on the evolutionary scale, the lower the enamel sodium. Within the horse family at least, enamel sodium content was a guide to stage of evolution.

I should clear up at this point a possible misunderstanding regarding how earth scientists work. Both geologists and evolutionary biologists are forced to deal with very complex "experiments" that have already been done for them. In contrast, in chemistry and physics scientists must grind through experiments to test their predictions. Stephen Jay Gould puts it well in a tribute to his Harvard colleague, biologist Ernst Mayer, when Mayer received the Balzan Prize. Gould wrote, "The Nobel prizes focus on quantitative, nonhistorical, deductively oriented fields with their methodology of perturbation by experiment and establishment of repeatable chains of relatively simple cause and effect. An entire set of disciplines, different though equal in scope and status ... is thereby ignored: the historical sciences, treating immensely complex and nonrepeatable events (and therefore eschewing prediction while seeking explanation for what has already happened) and using the methods of observation and comparison."

The point I want to emphasize is that geologists and their colleagues in related fields proceed according to the same sort of abrupt leaps that one finds in the fossil record. During research flashes of insight lead the worker in fruitful directions through the overwhelming mass of data that are available for study. The research is directed in an ad hoc manner by bursts of inspiration. In the course of Toots's and my work on sodium we first assumed that sodium levels were controlled metabolically and set out to test it. We didn't accumulate great masses of data and then propose our theories as would be the socially acceptable way for a chemist to do his or her work. Toots and I guessed at the outset that animals low in the evolutionary chain had higher body sodium levels, and we selected our samples for analysis to demonstrate our theory. The theory came first, not last. We chose to analyze a sequence of horses because the ancestry of the horse is well documented, thereby providing us with a quality test of our theory that sodium levels decreased with higher evolutionary stage.

When we looked at enamel from a variety of animals we found that there was indeed an overall pattern. Bison, contemporaries of *Equus* during the Pleistocene epoch (roughly the last two million years) which were successful enough to survive to today with few changes, had enamel sodium levels comparable to that of horses. Extinct lines of mammals such as the giant titanotheres (huge, knobby-headed relatives of the horses), which vanished millions of years before Pleistocene time, had markedly higher sodium levels than their more successful contemporaries, as did the proboscidians (the elephants and their kindred). We found later that carnivores had very much higher sodium levels than herbivores, with omnivores in between.

What did all our data mean? We hypothesized that the life sodium levels we estimated were related to evolutionary stage. If that were so, then the sodium pattern we perceived had to be consistent with evolutionary theories. Two such important theories are that marine animals are the ancestors of those that later lived on land and that lines of descent generally begin with carnivores, evolving toward omnivores and then herbivores. We expected that there should be a general trend of lower sodium with advances in evolutionary stage.

We found that our data fit the scheme. Enamel from a fossil animal similar to a sea-cow from marine rocks of the Tertiary age (the so-called Age of Mammals) in California showed the highest life sodium levels we found, 0.91 percent. The primitive titanotheres and an early rhinoceros called *Subhyracodon* had life levels of 0.84 and 0.70 percent respectively, whereas advanced herbivores like *Equus* and *Bison* gave 0.54 and 0.52 percent.

We have no data on fossil carnivores because they are rare in the fossil record and curators are reluctant to sacrifice their specimins for analysis. However, the enamel of modern coyotes contains 0.86 percent sodium and that of a modern bear (*Ursus americanus,* the black bear) contains 0.74 percent, fitting our model nicely.

To understand why all this should be so, consider that the diet of carnivores and marine mammals is relatively high in sodium, and hence that their physiological makeup must be designed to rid the tissues of excess sodium to maintain an optimum level. One would thus expect their tissues to be higher in sodium. Land plants are much lower in sodium than animal tissues, so an omnivore dining on them would have less need for sodium expulsion and hence less sodium in its tissues. Animals that ate only leafy plants (browsers) would take in still less dietary sodium and should have lower sodium levels than both carnivores and omnivores, as we find with the browsing titanotheres and rhinos. Finally, because grasses contain very little sodium, we would expect that not only would the tissues of grazers contain even lower sodium, as we observe in horses and bison, but that their physiology would be adapted to sodium conservation rather than excretion.

The Tertiary period, from about sixty-five to five million years ago, was a time when grasses appeared and expanded their range at the expense of leafy herbs, shrubs, and trees. Concurrently, this was also the time when grazing animals appeared to eat those grasses. The ancestors of the horse, the bison, and other ruminants were important and expanding lines. These animals were adapted to the low sodium diets that the grasses provided, and they prospered. Browsers such as the titanotheres were at a disadvantage in this setting because they probably required the higher intake of sodium that their diet of leaves provided in order to maintain high levels of sodium in their tissues, as reflected in their fossils. The extinction of the titanotheres and other lines with high sodium requirements may well reflect the lack of a suitable diet and a physiological inability to retain sodium in a world where the grasses were taking over the habitat of plants with high sodium content.

The extinction of the various elephantlike creatures such as the mastodon and the mammoth is probably at least partly attributable to the same cause. Both mastodons and mammoths have relatively high life sodium levels (0.68%) compared with their grazing contemporaries during the Pliocene and Pleistocene epochs (about five million to ten thousand years ago) and on the basis of sodium levels would be considered to be lower down on the evolutionary scale than animals like horses or bison. On the basis of their teeth, however, they would be considered to be highly advanced by most paleontologists, especially the wooly mammoth, which is closely related to modern elephants.

According to Toots's and my theory, elephants are primitive in their metabolism, and their ancestors should have been sodium dependent just like the titanotheres. This idea is borne out by studies done in Africa by wildlife biologists showing that the distribution of modern elephants is closely correlated with high environmental sodium levels and that they also select food that is unusually rich in sodium. One of their favorite foods is the bark of the acacia tree, which is richer in sodium than the average plant tissues and which they strip off to the extent of killing the trees in some cases.

It is not much of a step to imagine that the extinct mammoths of the northern hemisphere may have paved the way for their own extinction by killing trees for food both by barking them and also by breaking them down for feeding, another habit practiced by modern elephants. The destruction of trees and shrubs would have created more habitat for grasses and grazers. The grazers' habit of eating tree and shrub seedlings tends to perpetuate their food supply while assuring that the elephant's food supply does not renew itself readily. This is not to say that lack of sufficient sodium in their diet was the cause of the extinction of the mammoths in America (horses and camels, which had more advanced sodium metabolism, vanished at about the same time), but it could have been one factor that made them less competitive in the day-to-day contest for survival. Primitive sodium regulation may indicate that mammoths were poorly adapted in other ways as well.

Not only does geochemistry give us a more meaningful insight into evolution, but now we know why elephants at the circus or zoo like peanuts so much. It's not the nuts. It's the salt.

Chapter *8*

Girl with a Curl

*T*HE EXPLORATION OF OUR SOLAR SYSTEM WITH BOTH MANNED and unmanned vehicles has confirmed what anyone would have guessed—that the various planets and their moons are chemically unlike earth and unlike one another. We read with horror about the composition of the atmospheres of places like Venus or Jupiter because they sound more like something that would come in a laboratory bottle or a cylinder of compressed gas than like something that could be breathed. We can only speculate what the drinking water would be like in a place whose atmosphere contained things like methane, ammonia, sulfuric acid, or hydrochloric acid, chemical names that we associate with the pollution of our own world. Acid rain worriers, you haven't seen anything till you go to Venus.

But in all fairness to the Venusians or to any other inhabitants of any body in the universe, what about our own

For more information, see references 14 and 87 in *Further Reading.*

planet? How would it seem to a scientist from another world? Not very encouraging for habitation, I'd guess. Consider this planet we live on. Viewed objectively, even for humans it is not too promising. Most of the surface is covered with water. What part is land is not all a suitable habitat, to say the least. A great deal of the land is covered by ice and snow, some seasonally and some year round, and many of the areas that are not are too swampy during their so-called warm seasons. Still another portion of the land is bare rock or sand at high altitudes, terribly dry deserts, or locations where extreme winds make life tedious. The places that do provide conditions that are tolerable are subject to periodic flooding or droughts. Bleak though this may sound, to a visitor from outer space things would likely seem even worse than already indicated. If the alien forms were dryland dwellers they might be disturbed indeed to have periodic showers of a powerful solvent and highly reactive chemical compound—water—falling on them and their equipment. To make matters worse our atmosphere is a combination of mostly nitrogen, a relatively unreactive gas that would be an unlikely candidate for utilization by any life form, and a fiercely reactive substance, oxygen, that would combine, possibly violently, with both their tissues and their gear. Add to this the fact that the planet Earth is close enough to the sun to be bombarded daily by radiation intense enough to kill many earthly life forms in seconds and cause permanent tissue damage to higher animals such as farmers and sunbathers. Humans and many other organisms have become tolerant of these hostile environmental factors through selective survival over many hundreds of millions of years, but an alien would probably not last long, suggestions made by recent motion pictures to the contrary.

It might be assumed, though, that any space voyeurs who were clever enough to be able to scrutinize Earth vis-à-vis would have prepared themselves to cope with the highly disadvantageous conditions they would encounter by use of special clothing, breathing apparatus, and protection for their gear. Their biggest problem would be developing some local source of food unless their visit was to be a short one in which brown bagging would suffice. Let us suppose that the aliens were able to utilize such carbon compounds as lettuce, lamb, wheat, apples, and braunschweiger to provide for their table and larder. Could they then survive? Not very likely, unless their home planet happened to be a geochemical duplicate of ours in every detail. They would have problems not with the compounds of carbon combined with oxygen, hydrogen, and nitrogen but with the minor and trace elements that are found in all of our foods and water supply, elements like zinc, copper, manganese, selenium, arsenic, lead, and molybdenum. Perhaps even relatively abundant elements like calcium or phosphorus could be toxic to an organism that was not tolerant of them.

Life on earth has evolved in the presence of a host of chemical elements, over ninety of them, and has evolved to be tolerant of them in the amounts in which they are naturally available in the ambient geochemical environment. When a given element is present in amounts larger than usual, it is not infrequently poisonous. Yet many elements, even very minor ones, play important physiological roles in many organisms, and if any one element is not present in the usual amounts a deficiency disease may result. Every organism on earth has a specific tolerance for every element and may exhibit responses of a life-threatening nature to environmental departures from the "normal" levels of that element.

The environmental level of a given element can be wide-ranging depending on the specific environmental niche involved. A common element like sodium is present in vastly different amounts in different environments. Most land plants and many organisms that live in ordinary ponds and streams are highly intolerant of even moderate levels of sodium; pouring a bucketful of seawater onto a sodium-intolerant species such as your lawn grass could prove to be toxic. At the other extreme, there are many organisms that thrive in concentrated salt brines and would sicken or die if their sodium-rich world were diluted with fresh water from a heavy rain. Human beings develop serious conditions from an excess of dietary salt yet cannot survive without some sodium intake. There is a necessary balance between the needs of organisms and the supply of major elements such as sodium, calcium, potassium, iron, and others in common foods.

Many less common elements are also tolerated and even required by organisms, given their specific environmental levels. Looking at analyses of common rocks or of the crust as a whole (see the appendix), you might think that organisms would be adapted to high concentrations of aluminum, for it amounts to about 8 percent of most rocks. This is not the case, however; neither plants nor animals tolerate aluminum, even in relatively small quantities. This seems to be in conflict with the argument I have just presented, but it really is not. The point is that organisms develop a tolerance for the amount of an element *available* in the environment, that is, the amount that is able to enter into ordinary chemical reactions. The aluminum found in the earth's crust is not able to enter into the sort of reactions utilized by organisms. Aluminum forms highly stable, almost inert compounds with oxygen and hydrogen that tie it up so strongly that it cannot be readily assimilated and for all practical purposes might as well not be there at all. Because of the geochemical behavior of aluminum, its role in organisms is like that of a minor or trace element, and even relatively small amounts are poisonous. But don't walk into the kitchen and throw away all of your aluminum pots and pans. They are coated with an unreactive film of aluminum oxide and will not release aluminum in significant amounts to your food unless very acid or alkaline foods are cooked in them for a long time. A long-simmering spaghetti sauce or braised dish is unlikely to cause problems, but you could take the precaution of not using aluminum pots for cooking highly acid foods such as vinegar pickles or foods containing very alkaline solutions such as baking soda or lye, which is used to make hominy from dried corn. There are those who choose to avoid baking powder altogether because it contains aluminum, but there is little objective information to support the concept that trace amounts of aluminum from this source or from cookware is detrimental to health, although this is not universally agreed upon.

Another element that is not available in the amounts one might expect from observing average crustal compositions is copper. Copper amounts to about 50 parts per million (ppm) of the crust yet is present in seawater at only 0.001 to 0.025 ppm, and typical land plants contain about 5 ppm. The reason for the disparity between crustal abundance and the amount in water and organisms is that much of the copper of the world is combined with sulfur, and so it is not available for reactions that would put it into the food chain.

Copper is sufficiently available, however, that it is required by many organisms in small amounts. For example, mammals require copper for the manufacture of hemoglobin, the oxygen-carrying component of blood. It is therefore possible for a diet to be deficient in copper, although this is rare in humans. There are deficiencies of copper in some livestock diets, and copper supplements are provided in such cases. The copper requirements of different species are different, reflecting their different ancestries. Cattle require more copper than sheep, for example. Domestic cattle were developed from wild species living in Africa and Europe where environmental copper levels are about average or above, whereas sheep were domesticated from wild precursors that lived in Asia Minor in regions dominated by relatively low copper rocks like limestones and granites. We see the difference in the origins of the two species in their present-day copper tolerance.

Man also has a limited copper tolerance, a fact that has been known for a long time. The Swedish senate prohibited the use of copper cooking vessels by its armies and navy in 1753 because of the toxic effects of the metal, decreeing that all cooking pots be made of iron. The toxic nature of copper seems to be forgotten today, with copper being widely used in plumbing and expensive gourmet pots and pans. The inside of the pots and pans are generally coated with an inert layer of tin to eliminate any copper problems, as are copper vessels used to cook sausages or make beer in packing plants and breweries. Copper plumbing can have toxic long-term efffects, especially with water that is naturally corrosive because of its chemical composition. A geochemist friend of mine moved into an apartment house where the brand-new plumbing would have been especially reactive because of the fresh, clean surfaces and traces of soldering flux, and all of his tropical fish promptly sickened or died. I suggested that it was the high copper content of the new pipes, so he filled his tanks with water from another source, and the problem vanished.

The rarer element selenium takes its name from the moon and like the moon it has a bright side and a dark side, providing a particularly good example of an element that can be friend or foe depending on its concentration. Selenium is chemically similar to sulfur in some ways, though it does not commonly form insoluble selenides in the way that sulfur forms sulfides. Instead selenium forms mostly soluble selenites and selenates and thus is readily available chemically. Consequently, the biologically required level of selenium is roughly equal to its abundance in the crust, about 0.1 ppm. The fact that selenium is essential to organisms was not realized until just a few years ago because it was considered a toxic element, known as such since the days of alchemy.

Selenium is indeed toxic at about ten times (what scientists call an order of magnitude) its crustal abundance or above. Once again, livestock provide the best example. Selenium is present in higher than usual amounts in some shaly rocks of Mesozoic age in the Rocky Mountains, and the selenium in the soils derived from these rocks is further concentrated by various species of plants that have not only adapted to high amounts of selenium but thrive on it and concentrate it in their tissues, with some plants containing as much as 6,000 ppm of the element. Because of the harmful effects of selenium on cattle and sheep, a professor of agriculture at the University of Wyoming, O. A. Beath, studied seleniferous plants and their distribution for many years, imprinting the concept of selenium toxicity on the agriculture fraternity.

Thus it was a great surprise when some veterinarians in

the Pacific Northwest discovered what appeared to be adverse health effects from a *lack* of selenium. Their observation has been confirmed in many places for sheep and cattle, and the human health establishment has finally realized that selenium is an essential dietary element for man as well. Deficiency symptoms appear at about 0.01 ppm, or one-tenth the crustal abundance. Selenium therefore provides a superb example of a geochemical rule of thumb that Heinrich Toots and I published in the biomedical literature: You can estimate what levels of a minor element will be either toxic or produce a deficiency disease if you know the crutstal abundance and relative availability of the element to organisms. The available crustal level will be the required level for organisms, one-tenth that amount will produce deficiency symptoms, and ten times that amount will produce toxic effects. This rule is extremely useful but cannot be applied blindly because some very rare elements such as gold or platinum are not poisonous by virture of their chemical inactivity, and the same can be said for the noble gases in the atmosphere.

Some classes of elements seem to be toxic at almost any level, for example the so-called heavy metals such as lead or cadmium, present at levels of 15 and 0.2 ppm, respectively. As one would expect from their relative abundance, cadmium is more poisonous than lead, but neither is a very friendly element. The reason for their high toxicity is their being relatively unavailable because they are mostly combined in insoluble sulfides. Thus organisms have never had to contend with their presence and so have developed little tolerance for them. When people smelted lead ores (releasing cadmium as well as lead) and made pipes from the easily worked metal, toxicity began to show up. This was happening as early as the days of the Romans and contributed, some think, to the decline of the Roman Empire. To make matters worse for the Romans, the wealthy classes even added soluble lead compounds to their wine as flavoring, so the leaders were poisoning themselves at an even more rapid rate than was the man on the street. It is a commentary on the intellectual prowess of mankind that lead pipes were still in use only a few decades ago in many places.

The application of the rule of thumb for toxicity can again be illustrated by the element plutonium. The crustal abundance of plutonium is zero because this element is not a naturally occurring one, being a product of synthesis by various artificially induced nuclear reactions. (To be strictly accurate, there is said to be some naturally occurring plutonium in Africa where it formed in natural reactors in some uranium deposits at Oklo, Gabon, but it is essentially nonexistent in the earth as a whole.) If we apply the rule of thumb we see that a plutonium deficiency will only arise if there is less than none present. The tolerance for it will be at a level of zero, and any level above ten times zero will be toxic. In other words we could predict that it would be toxic at any level at all. This proves to be the case, and most authorities consider plutonium to be the most toxic of all elements. Plutonium offers the added feature that if you are not poisoned by it there is always the possibility of being killed by a nuclear blast from a plutonium-based device. Ah, the ingenuity of man.

All organisms, including humans, are of course remarkably adaptive, and many earthly creatures live in environments or have diets that would be toxic to most other inhabitants of the earth. There are even human examples, such as the arsenic eaters of southeastern Europe who consume the deadly stuff as a bodybuilding tonic at levels that would kill most of us. In fact, the name arsenic derives from the Greek word for strength or virility.

There are, however, limits to man's adaptation, and since we began to search out deposits of unusual elements and refine them, we have been poisoning our environment by opening the natural reservoirs, both physical and chemical, where trouble-causing elements have been stored. Many elements—mercury, for example—are now found throughout the environment as a result of mankind's actions. The results of unearthing mercury are twofold. First, man has mined the element, removing it from safe storage in the earth where it is mostly in the form of a biologically inert sulfide and where it would have been harmless except in very local situations. As a result, mercury vapor is now a pollutant in untold laboratories, workplaces, and dwellings. Worse yet, man has made mercury compounds that are biologically available and then dumped them. Dumping grounds have included Japan, where people have died or produced deformed children as a result of eating contaminated fish, and a variety of industrial sites such as paper mills, chemical plants, and at a government-contracted plant at Oak Ridge, Tennessee. Even seed grains treated with deadly mercury compounds used for fungicides have been eaten by uninformed people with dire results. Mankind will pay for carelessness with mercury for a long time into the future.

People and the familiar plants and animals of the surface of the earth have evolved to tolerate natural levels of all chemical elements. It doesn't take a geochemist to guess what tolerable levels will be for most elements. All you need to know is the element's crustal abundance and a bit about the natural compounds it forms and their biological availability. The effects of excessive manmade concentrations of chemical elements are all around us to see. It is up to us to have the collective wisdom to use the elements of our rocks for useful things without paying the price that the Roman aristocracy did or that children in old buildings with peeling, lead-based paint are still paying. There have been some praiseworthy programs for reducing lead poisoning in ghetto children, but it is too late for the Romans.

The elements of the earth are in many ways like the girl in the Mother Goose rhyme: "When she was good, she was very, very good, but when she was bad, she was horrid."

Chapter 9

As Fish in the Sea

*F*OSSILS HAVE BEEN RECOGNIZED AS THE REMAINS OF ANIMALS and plants for a very long time. While some people imagined that they were the tricks of a playful deity, most of our ancestors took a more realistic view and concluded that fossils had been left by extinct life forms and as such gave us a view of the geologic past. Even biologists and geologists differed as to whether fossils were evidence of successive acts of creation or whether they illustrated a sequence of evolutionary change, but everyone agreed that they were very useful for correlating—indicating the equivalence of one rock layer with another, as William Smith had done in England and Wales. Yet while geologists and paleontologists thought that fossils were good indicators of time, they had yet to understand fully that they are also good indicators of other

For more information, see references 19, 20, 76, 77, and 91 in *Further Reading*.

things. In assuming that all rocks that contained a coral-reef assemblage of fossils were time equivalent, the French paleontologist A. D. d'Orbingy missed the point, for in fact they all represented similar environments rather than similar antiquity.

Like any young science, paleontology went through stages of discovery, myth, description, classification, and utilization as a tool by other scientists (in this case, geologists) and finally matured into an independent field when its practitioners began to ask themselves what the fossils revealed about the history of life on earth. This is not to say that fossils are not still used as practical correlation tools, and there are still paleontologists who are mostly concerned with the description and classification of specimens, but many scholars now study fossils to learn more about the population dynamics of communities of extinct creatures as a basic part of reconstructing a picture of the past.

One bit of information that is hard to come by directly is the total number of animals that lived in the past. The problem is that you cannot collect all the fossils from a given rock unit because there is always some practical limit, such as time or money or accessibility. Even if such a collection were possible, the number of fossils in a given place would not bear any simple relation to the number of animals that once lived there except by setting some sort of minimum number on the ancient population. The minimum, however, is a somewhat useless number. The meaningful question is, How much larger was the real population? It is clear that we cannot return to ancient times and conduct a census, but we can compare ancient environments with modern ones and make some reasonable assumptions that will lend support to our estimates of past populations.

There are indications that the total number of living forms on earth has not changed much since the beginning of the fossil record. Steve Stanley, a Johns Hopkins paleontologist, has observed that the number of fossils in a given volume of sedimentary rock has remained about constant since the start of the Cambrian Period about 590 million years ago. Given that we can use the present as a guide to the past, we should first ask whether there are any really solid data about the earth's population at present. We find that there are supposedly accurate figures on the numbers of whales or pandas, but what about less spectacular species? For most plants or animals the numbers are very large, with phrases such as "all the tea in China" or "numerous as fish in the sea" hinting at the magnitude of the totals but giving nothing in the way of useful figures.

The direct way to determine the number of organisms is simply to count them. This is easily done with a flock of domestic sheep that can be run through a chute but is a great deal more difficult with a similarly sized band of mountain sheep scattered over a large area. It's even more difficult to count animals that cannot be observed easily, such as fish in a lake or ocean or small mammals in a field or forest, or larger animals that are widely scattered. Many workers—including Darwin, who studied earthworms as a model for other species—have dealt with the even more difficult problem of the population of invertebrates, but let's confine this discussion to vertebrates.

Although you can actually count each member of a flock of domestic sheep and get a true enumeration, most census figures are estimates because of the impossibility of counting each individual in the population, not to mention the possibility of counting some individuals more than once. The census of the United States is a good case in point. The Commerce Department insists that everyone must cooperate, and it has perfectly good reasons for advertising the penalties that exist for failure to comply: there are many who will intentionally avoid being counted and others, who are nomadic, who may not be in the right place at the right time or who may be counted twice by pollsters in two different parts of the country. The Census Bureau admits that its figures for inner city blacks and illegal aliens are low, and they use "fudge" factors to try to come closer to a true value. Any census is an estimate of the whole population from a sample, and the sample is by definition almost always smaller than the whole population.

There are census figures for many species such as game

animals that are of some use, and there have been attempts at enumerating some smaller species as well. University of Illinois ecologist V. E. Shelford gave the following estimates for North America:

ESTIMATES OF MAMMALS IN 10 SQUARE MILES OF UNDISTURBED DECIDUOUS FOREST IN EASTERN NORTH AMERICA

number of individuals	% of total
220,000 "mice"*	77.5
63,500 "squirrels"*	22.4
470 deer	<0.2
30 foxes	
5 bears	
2–3 pumas	<0.1
1–3 wolves	

ESTIMATES OF MAMMALS IN 10 SQUARE MILES OF UNDISTURBED PRAIRIE IN KANSAS

number of individuals	% of total
49,000 "mice"*	62.1
19,500 ground squirrels	24.7
10,000 shrews	12.7
300 bison	0.4
100 pronghorns	0.1

*"Mice" refers to mouse-sized animals including voles, small burrowing rodents and the like, and similarly for "squirrels."

In another study in Africa, L. H. Brown concluded that there were about 150,000 rodents and insectivores in a 12-

square-mile area around an eagle's nest, so the populations in Africa are of the same order of magnitude as in North America. Such estimates are in striking disagreement with the proportion of fossils of small animals to those of large animals found in American rocks of Tertiary age. The principal difference is that small mammals are usually not well represented in fossil collections. This is not to say that the small fossils are not present in the fossil deposit, but the tiny bones and teeth may have escaped the notice of collectors. Mike Voorhies of the University of Nebraska State Museum makes the eminently sensible comment that "since there are few modern communities in which small mammals, especially rodents, do not outnumber large ones, their absence or rarity in *any* fossil deposit should be viewed with extreme suspicion."

Census data for live communities cannot be extrapolated unchanged to the fossil record because the fossils preserve a record of dead animals, not live ones. The smaller mammals have shorter life-spans than the larger ones, so they should be expected to contribute more individuals to the fossil record than their larger contemporaries. Thus the proportion of potential fossils of small mammals in a fossil community should be two, three, or more times higher than the estimates of living populations, making the problem worse instead of better.

In real life, though, small animals are more likely to be consumed by scavengers when dead or simply removed entirely by predators that capture them alive. Thus many small mammals are removed from the population by becoming some other creatures' lunch. Such loss of potential fossils counters the tendency for the over representation of small animals in the fossil population caused by their rapid reproduction and short life-span. As a rule, then, small fossils (which should include most birds as well) are likely to be underrepresented in a collection of fossils because of hungry contemporaries, a tendency to be washed away easily, and just plain sloppy collecting.

It is instructive to look at the opposite end of the size scale as well. Large fossils are not only easier to find but most likely will actually be overrepresented in a given deposit because they are more massive and therefore more resistant to destruction by chewing, decay, and transportation processes. Many a toothy skull or large fossil mollusc adorns a porch in the plains states because they are so exceptional in aspect that anyone who passes by, even an amateur, is likely to notice them and bring them home as a souvenir. Perhaps the most obvious fossils in the plains are those of the mammoth and its relatives. Owing to their great size, the skulls and bones of mammoths are discovered with regularity in excavations for houses, gravel pits, road cuts, and other man-made exposures. Mike Voorhies has calculated that approximately one out of every thousand mammoths that ever lived in Nebraska has been discovered as a fossil. The data are not as readily available for mouse-sized animals that lived in Nebraska in the same time period, but it is likely that museums contain not many more fossils of small mammals than of elephants even though the *potential* contributions of the former to the fossil record must have been millions of times higher. The turning point in collecting techniques and emphasis in vertebrate paleontology came when screening was introduced to separate the small fossils from the matrix material. The big game hunters retired and the work was passed on to the lovers of miniatures. This transition took place in the 1930s at institutions such as the Universities of Michigan and Kansas, where workers started washing tons of unconsolidated matrix through wire meshes. At the University of Nebraska, whose collection of fossil mammals displayed in an outstanding public museum boasts huge trophies, screening techniques were not utilized until the 1960s.

Modern censuses and estimates are probably a much better guide to the size and proportions of ancient populations than the fossils themselves because we can use them to construct theoretical models. Paleontologists and biologically inclined scientists have plunged into calculations of animal populations during the past decade or so as the life sciences have grown out of a descriptive phase into a predictive one.

Calculations can be very simple or highly involved, but

most of them are instructive and even fun. For example, I made the statement above that there were undoubtedly millions of times more mouse-sized mammals in Nebraska than there were mammoths during the same time period. This number can be arrived at by assuming that the total weight, or biomass, of "mice" was about equivalent to the total weight of mammoths. This assumption is a reasonable approximation in view of census data from many modern environments. An average mouse is outweighed by a mammoth by a factor of about 200,000. Thus, at any given time there should be 200,000 "mice" for every mammoth. Consider, though, that elephants live about fifty years and mice about six months, so that there are 100 times as many "mice" available for fossilization than their trunked colleagues. Thus there would be 20,000,000 "mouse" corpses for every mammoth corpse if the biomass ratio was 1:1. You should remember that an unknown, though presumably large, proportion of the "mouse" carcasses would have been consumed by scavengers of various sorts and thus had failed to enter the fossil record. Let's assume that only 1 percent of them were preserved. According to the collection records at the University of Nebraska, as of the 1930s there were fewer than 10 "mouse" fossils whereas there were over 1,000 fossils of proboscidians. This means that only 1 "mouse" skeleton had been found, or at least collected, for every 200,000,000,000 mice that lived in the state during the late Pliocene and Pleistocene epochs.

To illustrate how things have changed, Mike Voorhies and his colleague George Corner excavated a Pleistocene site (about one million years old) at McCook, Nebraska, during the summer of 1983. They recovered 3 proboscidians (elephants and their kindred) and over 5,000 "mice." This was in spite of the fact that they used ordinary window screen instead of a more satisfactory sieve and only screened the most promising 10 percent of the material. The ratio of 3:5,000 is a far cry from the 1:20,000,000 estimated above, but if Voorhies and Corner had found the same total number of mice in the remaining "less promising" 90 percent of the material not sieved, then the ratio of probiscidians to mice would have been 3:10,000. From my estimate that only 1 percent of the "mice" had been preserved you would expect a ratio of 3:600,000. Either Mike and George lost a lot of fossils through their screen or my estimate is incorrect in some way.

I made the assumption that the total mass of "mice" was equivalent to the total mass of "elephants." This assumption is reasonable in some environments, especially marine ones where it has been shown that biomass is uniformly distributed relative to the log of the weight of individuals. This means that animals in the weight range 1 to 10 grams should have the same total weight (the sum of *all* of the individuals in the range) as the total of animals in the range 10 to 100 grams, 100 to 1,000 grams, 1,000 to 10,000 grams, and so forth. This rule of thumb does not work quite as well with terrestrial communities, especially with animals of large size. In the tables given earlier the total small animal mass is only half that of larger species for the forest environment and only about a fiftieth for the prairie setting. If you assume a ratio of total biomass of 1:50 for "mice":"elephants," then the predicted ratio for Voorhies and Corner's collection would be 3:11,000—and that's pretty close to the probable 3:10,000 ratio for their site. Paleontologists are clearly now collecting more representative samples than in the past.

Other much more complex calculations are made by people concerned with the population of past and modern organisms. One general type of calculation depends on knowing the available energy inputs to an area and then calculating how much biomass it could support. For example, you could start with the total solar energy transfer, or flux, per unit area over the course of a year and then apply corrections for the efficiency of photosynthesis by the flora of a region. From these quantities you could calculate an estimate of how many animals the region could support because the efficiency of feed conversion of many species, prey:predator ratios, and other necessary quantities are known or can be approximated. Such calculations require a few pieces of data that are usually available and can be used to set rude limits on possible populations of regions that cannot be evaluated in another way.

Two Canadian marine biologists, R. W. Sheldon and S. R. Kerr, tackled the question of the population of monsters in Loch Ness, and their treatment, however tongue in cheek, is an instructive example of a theoretical approach to population estimates. The authors assumed that there are monsters in Loch Ness because they have been reported over many centuries. Sheldon and Kerr noted that the known characteristics of Loch Ness monsters are that they are rarely seen and never caught. The fact that they are rarely seen suggests very strongly that the population is small. In addition, when seen the monsters are invariably described as large, again suggesting that the population is small simply because the monsters are big and hence it would take relatively few of them to make up a large total biomass.

Sheldon has shown in a previous publication that whereas the production rate of oceanic organisms is size-dependent, the quantity (total biomass) of organisms is constant at all sizes. In other words, taken over suitable size intervals the mass of monsters in Loch Ness should be the same as that of fish, or plankton. This is the same sort of argument made for the mass of "mice" in Nebraska being roughly the same as that of "elephants."

Sheldon and Kerr couldn't locate any hard data about fish production from Loch Ness, but by comparing it with grossly similar lakes they estimated that the annual fish yield should be less than about 1 kilogram per hectare (0.9 pounds/acre). They refined this figure by using some data from Loch Lomond that would suggest that the yield would be about 0.55 kilograms/hectare (0.5 pounds/acre). Experience in other areas shows that total biomass ranges from equal to up to about five times the annual production. Therefore the total biomass, or standing stock, of fish in Loch Ness should be in the range 0.55 to 2.75 kilograms/hectare (0.5 to 2.4 pounds/ acre), and the total mass of monsters should be close to that amount.

Loch Ness has an area of 5,700 hectares (14,085 acres), so

the total mass of monsters can be inferred to be in the range 3,100 to 16,000 kilograms (6,900 to 34,000 pounds) in round numbers. Sheldon and Kerr stated that the minimum average size of Loch Ness monsters must be at least 100 kilograms (220 pounds) because ". . . anything smaller is not suitably monstrous." I think that most of us would agree that they are being very conservative in this regard since any proper monster should weigh at least a ton or so, but even allowing for a monster as small as 100 kilograms means that there are something between 1 and 160 monsters in Loch Ness.

The existence of only one monster would mean that not only could it be as large as 16,000 kilograms, but it would have to be very long-lived, for there have been sightings reported from Loch Ness for hundreds of years that we know of. A more defensible lower limit of the number of monsters would be two individuals, which would provide a minimum breeding population for sexual reproduction. (According to Charles Cole of the American Museum of Natural History, there are female whip-tailed lizards in the American southwest that are able to reproduce asexually so we cannot exclude that possibility with Loch Ness monsters, although such a mode of reproduction is *very* rare in vertebrates.) This minimum is still unlikely for a number of reasons. The most obvious one is that there is a 50 percent chance that two monsters would be the same sex and therefore incapable of reproduction. Also, if the pair were capable of reproduction, there would surely be three or more monsters in short order. With very large monsters one could also reasonably assume that they were very slow to reach sexual maturity, so that the total number of monsters probably consists of at least two of the opposite sex, and a number of sexually immature individuals. Thus the practical minimum number must be larger than the theoretically possible minimum. Without going into all the detail that I have, Sheldon and Kerr gave an educated guess of ten as the smallest viable population, a reasonable-sounding estimate.

They made a sensible assumption that the monsters are

probably at the end of the food chain in the lake and subsist largely on fish. The efficiency of feed conversion of many aquatic predators is cited by the authors as about 10 percent. If Loch Ness monsters are similar, then, based on the assumed fish production of the loch, monsters would increase at the rate of about 300 kilograms/year or more. If we were to assume that the monsters were small (100 kilograms, or 220 pounds, minimum), then one would expect that on the average three of them would die each year to balance off the 300-kilogram annual increase in monster biomass. Larger monsters would of necessity die less frequently on the average.

Two lines of evidence support the assumption of large size. First, corpses are never found, and we would expect an occasional one to wash ashore if as many as three monsters a year were expiring, or even as few as one. Second, as Sheldon and Kerr pointed out, if there were a high adult mortality, then there must be a big population of young monsters to replace the dead adults if the population were to be maintained. Young (small) monsters have never been reported in Loch Ness.

The two scientists concluded that the various factors favor monsters that weigh about 1,500 kilograms (3,300 pounds), which would mean that Loch Ness could support about ten to twenty of them. They noted that a 1,500-kilogram monster would be about 8 meters (26 feet) long, which is in general agreement with reported sightings. After taking into account possible errors, which would tend to cancel one another in any case, they concluded that their estimate of population density for Loch Ness monsters is close to the true value.

Their final conclusion was that many monster sightings, in Loch Ness and elsewhere, go unreported and that

Fear of ridicule is the main reason why many observers do not make their observations known to science. But it is the skeptics who are at fault. Monster observers should be encouraged. The occurrence of monsters is quite reasonable and is by no means fantastic. [Until brought to

our attention] . . . we were unaware that monsters were a problem.

At the conclusion of this short paper any reader who has become convinced of the existence of monsters in Loch Ness should glance back at the first sentence of the work—"It is well known that there are monsters in Loch Ness"—and realize that the title is *The Population Density of Monsters in Loch Ness.* All of the reasonable arguments that Sheldon, Kerr and I have presented relate only to the *number* of monsters in Loch Ness, *not* to whether or not they exist at all. While it is fair for Sheldon and Kerr to state as they did in the passage quoted above that the occurrence of monsters is reasonable, that does not mean that they do, in fact, occur.

The principles that Sheldon and Kerr used in their witty paper are derived from both observation, including censusing, and from theoretical considerations. At the root of any population estimate is the total energy input into the system as it controls the production of food at the bottom of the food chain. For a terrestrial environment the solar energy flux is a factor that can be applied more or less directly, more so than in a marine situation, where much of the energy is stored in heated water or in the form of organisms or nutrients that may be swept vast distances and to great depths and back prior to entering the food chain.

Given an ecological system with a certain amount of energy provided to it, a predictable number of living things can be supported. Sheldon and Kerr extrapolated data from other environments to make their estimates about Loch Ness life forms. Given a population of fossils, the paleontologist can make what appear to be reasonable estimates of past populations. Increased understanding of modern, living communities and more careful, sophisticated methods of collecting fossils will improve the hindsight of paleobiologists. Until someone invents the time machine of science fiction their calculations will have to stand.

Chapter *10*

A Fossil's Fare

O NE OF THE FUNDAMENTAL PRINCIPLES THAT IS TAUGHT TO architecture students is that form follows function, sometimes called honesty of materials. The idea is that a building will be more pleasing in appearance if the materials used serve both a structural and an aesthetic function. By this reasoning the columns of the Parthenon are attractive and some great, overblown columns in front of a university library are not. The principle can be applied to a whole building, too, with the idea that a bank should look like a bank, an outhouse like an outhouse, and a gas station like a gas station. I'm not sure whether this last idea is universally applicable, but I do think that banks that resemble teepees or gas stations that simulate old English cottages look pretty tacky. The principle is overdone in places like southern California,

For more information, see references 14, 15, 57, 59, 62, 69, 71, 72, 85, 86, 88, and 94 in *Further Reading*.

where a hotdog stand might look like a frankfurter on a bun, complete with mustard, or a coffee shop like a cup of coffee, but at least these structures serve the function of identifying a product, which cannot be said for the English cottage gas station.

With living organisms we can use a reverse version of the principle and state that function is revealed in form. Thus light-limbed quadrupeds such as horses, gazelles, and cheetahs are fast runners; animals with large claws on front limbs are likely to be diggers; and creatures with fins are undoubtedly swimmers. Such inferences with living forms are easily verified, but those regarding most extinct animals are not. Paleontologists use comparative anatomy to draw conclusions about animals long dead by using the structure of their skeletons to infer their life habits. A dinosaur with short, sturdy legs and a large body was likely a slow mover whereas one with large rear limbs and diminutive front ones was probably a high-speed, bipedal type.

A similar interpretation of the form of fossils appears in *Geology and Mineralogy* (1837) by William Buckland, who was Reader in Geology and Mineralogy and Canon of Christchurch College, Oxford. Referring to the fossil mammals of the Pliocene Epoch of England and Europe, Buckland reported that ". . . the individuals of every species were constructed in a manner fitting each to its own enjoyment of the pleasures of existence, and placing it in due and useful relations to the animal and vegetable kingdoms by which it was surrounded." Rev. Buckland's religious view of a fossil assemblage contains the same concept of function revealed in form.

The diet of extinct animals is also an area in which comparison with living forms is used. Often the whole body can be taken as a clue to dietary habits. For example, we may assume that a very large animal was probably a herbivore because it would not have been able to move quickly enough to catch enough animal prey to survive. More often, though, it is the structure of the skull, especially the teeth, that is used as the principal evidence for dietary practices. Animals with pointed, shearing teeth are assumed to have been flesh eaters and those with low-profile, supposedly grinding teeth vegetarians. It is unlikely that paleontologists have made really gross errors by guessing diet from the structure of an animal's teeth, but it is unnerving to hear that a favorite food of coyotes is watermelon or to see supposedly herbivorous ground squirrels eating their fellows' squashed remains on a highway.

There is a small amount of direct evidence of dietary habits in the fossil record. Such evidence includes the tooth marks of swimming reptiles imprinted into the shells of the snail-shaped molluscs called ammonites found in the Cretaceous seas a hundred million years ago and the tooth marks of crocodiles in turtle shells found in lake sediments of Tertiary age in Wyoming. Mike Voorhies of the University of Nebraska State Museum has found "seeds" of grasses in both the mouths and the body cavities of extinct rhinos. Frozen mammoths have been found in Alaska and the Soviet Union with stomach contents intact. In caves in the Nevada desserts, paleontologists have even found the dung of the giant sloth *Nothrotherium*, which resembles giant-sized horse droppings with partially digested remanents of meals, preserved by the dryness like so many Dead Sea scrolls. Likewise the fossil scat, called coprolites, of both aqueous and land-dwelling carnivores have been found, preserved because of the relative stability of their components: fragments of bone, teeth, and hair. Such finds are rare enough, however, to provide relatively little information about the diets of extinct animals.

Alan Walker and others at Johns Hopkins University have approached the problem differently by using scanning electron microscopy to discern small surface features of modern and fossil teeth. He maintains that he can distinguish the wear produced by browsing on soft leaves from that produced by grazing on abrasive, siliceous grasses.

Methods that are based on the physical properties of skeletal elements, particularly teeth, can only be applied to fossils of unusual perfection in which the features of interest are sufficiently well preserved. Like such direct evidence as coprolites or stomach contents, such fossils are not at all common. It would be highly desirable to have a method for making inferences about diet that was more broadly applicable.

When Heinrich Toots, presently teaching what some traditionalists might consider the unlikely pair of paleontology and geochemistry at C. W. Post College, was a graduate student at the University of Wyoming, he analyzed a large number of vertebrate fossils for a number of minor elements. His initial study was a survey of the fossils that were more familiar to him using his newly found skill at X-ray spectrometry. This explorational phase of study was appropriate at the time because little was known about what minor and trace elements were found in vertebrate fossils. Toots suspected from his preliminary data that of all the various elements in the fossils, strontium was the one that was most likely to be introduced through the animal's diet.

This conclusion was plausible for a couple of reasons. First, it was known that strontium could proxy for calcium in the mineral apatite, which makes up the mineral portion of vertebrate skeletons, and there had been a flurry of work on the matter by the biomedical fraternity in the 1950s during a period of concern that radioactive strontium from atmospheric nuclear testing would become incorporated into human bone. (For more about bone chemistry and mineralogy, see Chapter 7.)

Previous studies of strontium had shown that the amount that enters an animal's tissues does not appear to be regulated in any physiological way, and while there is some discrimination against strontium in favor of calcium in skeletal tissue, D. L. Thurber and others had previously shown that the levels in bone generally reflect intake. In addition, the fact that within a single skeleton the strontium level was approximately constant suggested that the strontium in bone was not readily removed or added by postmortem processes. Otherwise, it would have been more stable in dense, impermeable skeletal elements such as teeth than in open, permeable portions such as spongy bone.

Thus strontium levels in vertebrate fossils appeared to

hold promise for making inferences about diet and environment, at least for land-dwelling animals. (Strontium levels in
aquatic vertebrates are determined largely by environmental
water composition, according to physiologist H. L. Rosenthal
of the Washington University [St. Louis] School of Dentist-
ry.) Environmental levels of strontium have a considerable
effect on skeletal levels in terrestrial animals. For example,
Toots and I studied a large suite of modern samples from the
rift valley in Kenya that were collected for us by Kay Behrensmeyer (Anna K. Behrensmeyer, who is now with the
Smithsonian) in which we found high strontium levels relative to other samples that we had examined, reflecting the
high ambient strontium levels in that region. It was clear that
if the effect of diet was to be examined, then all the fossils
analyzed must have lived together in the same environment,
at least with respect to environmental levels of strontium.

It is fortunate that Mike Voorhies was a Wyoming graduate student at the time who had also taken my course in geochemical analytical methods. He and Toots teamed up to test
the relation between diet and strontium. At the time Mike
was excavating a vertebrate fossil site near Orchard,
Nebraska (a site he referred to as Verdigre NE) as part of his
Ph.D. dissertation research. The site was not only rich in fossils but apparently catastrophic in origin, for a large number
of coexisting animals had been killed and buried together.
Thus the fossils were from animals that had shared a limited
life environment together and had presumably been exposed
to comparable geochemical background levels. For these reasons, the site was almost ideal for testing the strontium–diet
relationship because all other factors were constant. This
being the case, any differences in the strontium content of
their bones would be the result of differences in diet. In their
first studies Toots and Voorhies confirmed that strontium
levels were the same for different parts of the skeleton of a
single individual, suggesting both that the strontium was
from dietary sources and that it remained stable after the
death and burial of the animal.

From previous studies, plants were known to contain different levels of strontium. In general, grasses contain less of the element than leaves of woody plants because grasses apparently excrete strontium in dewdrops whereas leafy herbs and trees tend to retain it. Thus grass eaters (grazers) should ingest less strontium than leaf eaters (browsers). Food-chain effects also alter the strontium levels found in certain animals. Because herbivores concentrate strontium in skeletal tissue, carnivores would be expected to derive a smaller amount of strontium from dietary sources. Carnivores that eat bones derive little from the mineral portion of the bone, where strontium is concentrated, as evidenced by the fact that carnivore feces are almost entirely made of bone fragments. This is not to say that bone eaters would not acquire some strontium from the skeletal tissues, and you should be able to distinguish bone eaters from carnivores that eat flesh only, although nobody has tried this as far as I know.

Toots and Voorhies selected fossils from carnivores, two grazing mammals (a horse, *Protohippus*, and a proghorn-like mammal, *Merycodus*), and two browsing animals (a horse, *Hypohippus*, and a turtle, *Testudo*). The analytical results are shown below.

STRONTIUM CONTENT OF FOSSIL BONE

bone source	number of specimens	strontium (ppm)
carnivores	4	477
Merycodus	30	526
Protohippus	10	552
Hypohippus	10	630
Testudo	10	636

The bone strontium levels were clearly a reflection of diet, and as predicted, the carnivore samples had lower levels than any herbivore. Among the herbivores the grazers' bones contained less strontium than the browsers'. Here, then, was a powerful tool for determining the diet of extinct animals. One especially attractive thing about this method is that any skeletal tissue can be used whereas teeth were required, generally speaking, to make a dietary judgment based on morphology alone.

As one might expect, strontium alone does not provide *all* the answers, but together with other information such as dental morphology it is a useful tool. Bone from *Diceratherium* (a primitive three-toed rhinoceros with two side-by-side horns at the tip of its nose) from Agate, Nebraska, contains 0.16 percent strontium, whereas that from the associated *Moropus* contains 0.12 percent (there is no overlap in the strontium values for the two forms). The combination of a high strontium level plus the dental morphology suggests that *Diceratherium* was a browser, a reasonable view in terms of its large size as well. *Moropus*, a large, clawed relative of horses whose habits are uncertain, has teeth that eliminate any possibility of it being either a grazer or a carnivore. The low strontium content of its bones lends support to the speculation that *Moropus* was a digger that subsisted largely on roots and tubers.

The use of strontium as a dietary indicator was accepted quietly by the paleontologic profession but welcomed with enthusiasm by anthropologists. Primates, including man, have dietary habits that bear little relation in many cases to dentition. The gorilla, with its prominent canines, is a strict vegetarian, and man, who possesses generally noncarnivore teeth, is an omnivore, though generally a carnivore if food supply permits. The diet of early man and man's precursor hominids was a major unsolved mystery, and any tool was seen as valuable.

Anthropologists all over the world started analyzing bone samples with variable results. Antoinette Brown, an anthropologist at the University of Florida, tested some American Indian remains and concluded that she could distinguish both sexual and social differences among the group. The lower social classes ate less meat, although the males among the lower class consumed more than the females. Nobody would question her results today, but at the time some anthropologists were critical of Brown's conclusions, although the samples she analyzed all came from a single burial site and her interpretation was reasonable. New ideas are always accepted slowly.

Toots and Voorhies cautioned in their original paper that the specimens analyzed must come from a single site, a single quarry, if any comparisons are to be made, for this would make it more likely that environmental strontium levels were the same for all the living animals represented. A number of anthropologists succumbed to the temptation to ignore this advice and obtained inconsistent results, which of course they blamed on the method rather than on their application of it. The method continues to be used and has provided significant results for those who use it properly. However, there are drawbacks to applying any chemical method to human or hominid remains: the scarcity of bone for analysis and the fact that you must generally have a skull or a jaw to identify the fossil—and skulls and jaws are mighty rare.

Margaret Schoeninger of Johns Hopkins is one of the more successful practitioners of the paleodietary arts, having tried a great variety of analytical methods and guarding carefully against errors in assumptions or collections. Her dietary reconstruction of an archeological site in Mexico is a classic in the field. She is presently using the ratio of two stable (nonradioactive) isotopes of nitrogen to determine the relative contributions made by marine and terrestrial food sources. Marine plants have more of the heavier N^{15} than terrestrial plants and the resulting higher $N^{15}:N^{14}$ ratio persists throughout the food chain. The nitrogen that persists in fossils in bone collagen reflects the ratios of the isotopes in diet and serves to distinguish fisher-gatherers from agriculturists. She and others are also using ratios of stable carbon isotopes in a similar way. A group of paleodietary researchers is meeting for an exchange of ideas in California in early 1984, so this young and esoteric field is undergoing an expansion that promises to provide all of us with more information.

I suppose it is reasonable to assume that future scientists will have a complete kit of tools to distinguish those among us who favor the cheeseburger over the burrito or a steak over tofu with garbanzos. And if you are one of those who sneaks a chocolate chip cookie now and then and who thinks that you're getting away with it—forget it. It's in your bones.

Chapter 11

Gold Is Where You Find It

*A*S A YOUNGSTER GROWING UP IN SOUTHERN CALIFORNIA I was always fascinated by mines and mining. My father and I used to visit active mines in usually vain attempts to get a guided tour, and there was nothing quite like prowling around the abandoned tunnels, headframes, and mills that dotted the deserts. Every claim notice we located was read carefully and then replaced in the Prince Albert tobacco can mounted on the claim post—upside down of course, so as not to accumulate rain.

If I couldn't be in the desert, I'd peer into the assayer's shop on Hollywood Boulevard and even got up the nerve to talk to the busy, and somewhat grouchy, assayer a few times. There was also an old copper mine with a U-shaped tunnel dug into the brown rocks of the Hollywood Hills that was

For more information, see references 1, 26, 50, 78, and 97 in *Further Reading*.

126

only a short bike ride from home, and nearby was a wonderful abandoned basalt quarry with tunnels to explore and minerals to collect when some movie wasn't being shot there. I recall spoiling the movie *Julius Caesar* for the geology librarian at Berkeley when I told her, years later, that in one dramatic scene she could see the buildings of downtown Hollywood through the smog as one of Caesar's legions marched through my old, familiar quarry. Unlike the other kids, who rooted for the cowboys in the Saturday matinee Westerns, it was the prospector who was my hero.

I think the main reason that I was so intrigued by the prospector was that it was not at all obvious to me how the old buzzard, usually played by the best character actor that the studio could muster, located a mine site. In *Treasure of the Sierra Madre* how could Walter Huston simply walk off into the Mexican mountains and locate a gold mine? When I looked at the rock around prospects and abandoned mines I practically never saw anything that looked like ore to me. I chalked that up to my own ignorance and continued collecting and visiting museums. Eventually I ended up studying geology at Berkeley.

I took all the economic geology and mineralogy courses that were offered, partly so that I could become a professional geologist but also to satisfy my youthful curiosity about how prospectors located mines. Naturally I learned quite a bit about ore deposits, but what I learned was about ore deposits that had already been found and exploited. Economic geology is largely taught through descriptions of case histories of the geological relations and development of specific mines. I concluded as an undergraduate, and have found nothing in the several decades since to change my opinion, that mines are not located by scientific means, or even by scientists. Almost all mines, from the Bronze Age to the present, have been located almost accidentally by persons other than professional geologists.

There have been mining geologists since the time of the great Georgius Agricola (known to his fellow Saxons as plain Georg Bauer) in sixteenth-century Saxony, and doubtless before then, but even the leaders in the field seem to have been distinguished mainly by their excellent theories rather than by their skill at locating new deposits. It seems that despite theories regarding the formation of ore deposits, gold is where you find it.

Of course, locating ore deposits was considerably easier in 9000 B.C., as the Western world eased into the Bronze Age, than it is now. Now there is probably no place in the world where one could expect to stumble on a deposit that was so rich that it could be recognized on sight.

We can only guess at what mining was like ten thousand years ago, although some archaeologists have studied ancient workings. The first mention of ore deposits in Western literature was made by Aristotle, who thought that ores were the result of some sort of emanation from the stars. For almost two millennia, essentially nothing was written about the origin of ore deposits. Then in the fifteenth century the alchemists appraised the problem with the apparent consensus that ores came somehow from heat within the earth. The source of the heat was unknown, but Aristotle's idea had not lost its punch, as seen by the following.

> For cause sufficient Mettals finde ye shall
> Only to be the vertue Minerall
> Which in every Erth is not found,
> But in certain places of eligible ground:
> Into which places the Heavenly Sphaere,
> Sendith his beams directly everie yeare
> And as the matters there disposed be
> Such Mettalls thereof formed shall you see
> Ordinall of Alchimy, THOMAS NORTON, 1477

Norton expressed the generally held view that the source of the energy was from the heavens, a logical assumption since almost everyone thought that the earth was the center of the universe, and where else should all the mystical rays converge if not the center? Norton also notes that the ores are only found "... in certain places of eligible ground ... ," indicating that it was understood that the geologic structure or rock type in some way determined where ores were likely to be deposited. It is clear from the examination of ancient workings that the miners understood that the ores were localized by the geological properties of the enclosing rocks. The view that the stars and planets were the source of the heat that produced ore deposits lingered for many centuries. In fact, historians think that in 1790 Spain readily renounced any claims it had over British Columbia because it was assumed that the area was too far north to have been bathed with sufficient sunlight to produce valuable ore deposits.

Theories improved, or at least changed, with time. It was recognized that some deposits were vein-type deposits and that others were confined to certain sedimentary layers. Contrasting theories of origin were proposed, one that the metals were dissolved out of ordinary rock and redeposited as ores and the other that ores were deposited by waters emanating from cooling igneous rocks. Some thought that veins originated at the surface and others that they came from the earth's depths. Some of these arguments have gone on to this day. I don't think that I'm being unfair to the economic geology community if I state that since the eighteenth century its task has generally been to refine old theories rather than to develop any really new ones. As I will discuss later, a test of these more modern, fine-tuned theories is coming up in the immediate future.

Theories of ore deposition and localization all seek to explain why large quantities of a given element or suite of elements are found in some places and not others. One simple theory is that the earth was originally inhomogeneous and that we now find regions containing a number of deposits of a given type because that chunk of the original material of the earth was rich in certain metals or elements to begin with. This explanation is difficult to support if we accept the concept that the earth was molten at some time because in such a state mixing should have been extensive. Even from a cold start, the earth likely reached very high temperatures during its long evolution. Another theory is that an inhomogeneous crust was formed during Precambrian time, some billions of years ago, by the accretion of large meteorites. Suffice it to say that most geologists do not accept the concept of an inhomogeneous crust as a major explanation of the location of ore deposits.

That being the case, a theory is needed to explain how elements become concentrated in economically minable quantities in a given locality or region. By what natural process does a minor element in the crust become enriched by many orders of magnitude over its average crustal abundance? For example, very rich gold deposits contain several ounces of gold per ton of ore, or approximately 0.01 to 0.02 percent gold. The average crustal abundance of gold is only 0.05 ppm, or 0.000005 percent, so such a gold deposit would contain several thousand times the amount of gold in an average piece of the crust.

Some very rich deposits result when less valuable components of the original deposit are removed, leaving behind the more valuable residual material. One common mechanism for this is the weathering and removal in solution of unstable minerals. Gold, platinum, and many precious gems are highly resistant to solution and are thus concentrated in such residual deposits. In addition, transportation by water may sweep away fragments of low density and leave the same chemically resistant materials still further concentrated in what are called placer deposits. The gold mines of pre-Columbian South and Central America, the California and Alaskan gold rushes, Erskine Caldwell's *God's Little Acre*, and probably King Solomon's mines in Ophir were of these types—rich, easily found, and easily worked with simple equipment.

By the mid-nineteenth century, and even much earlier, it was generally agreed that ore deposits that were not obviously of sedimentary or residual origin were formed by the deposition of unusual concentrations of elements from some type of fluid, and some even thought that placers formed in that way. These fluids might deposit their dissolved matter as veins, or the matter might be disseminated through rock, or there might be some other arrangement, but few doubted the existence of the fluids. It had also been noted for centuries that many ore deposits were associated with volcanic or other igneous activity. Many geologists combined these two concepts—an association with igneous activity and the need for an "ore-forming fluid"—and concluded that the igneous mass was the source of the fluid. Different geologists proposed slightly different theories, but common to them all was the concept that many unusual elements such as gold, copper, silver, mercury, and others were not able to fit into the crystals of common silicate minerals that grew during the cooling of an igneous mass. Therefore the remaining liquid in a slowly crystallizing body of igneous rock would have become richer and richer in these rare elements, thereby forming the ore-forming fluid. To support the theory, geologists pointed to the fact that ore minerals were presently being deposited around volcanic vents, hot springs, and the vents of high-temperature gas called fumaroles. The fluid was thought by various workers to be acid, alkaline, saline, liquid, gaseous, and so forth, but despite these arguments over the details of its chemical makeup, there was general agreement among one segment of economic geologists that the fluid was of igneous origin.

In natural science there is always difference of opinion, and as one might expect, economic geologists did not all agree. Another camp pointed to many deposits that showed no known association with igneous rocks and yet appeared to have been formed by deposition from an ore-forming fluid. These geologists insisted that the metals in the fluid were derived by leaching from ordinary rock. This concept goes clear back to Agricola and probably before his time but was first formalized as the theory of lateral secretion by Sandberger of Würtzburg, Germany, in 1882, following work done by G. Forchammer of Copenhagen in 1835. Lateral secretionists believed that as it moved through common rocks, ordinary groundwater could dissolve elements that were present in them in small amounts and then act as an ore-forming fluid, redepositing the elements to form an ore body in some other place.

The igneous camp allowed that some deposits that were in

no known way associated with igneous activity might have arisen through a lateral secretion mechanism but insisted that most ore deposits had an igneous source. The secretionists continued to maintain that the igneous rocks associated with ore deposits were merely a source of heat and did not provide the ore elements themselves.

One major difficulty that the igneous camp had to overcome was the fact that most ore minerals, mainly sulfides of metals, are highly insoluble in water, the principal component of the hypothetical ore-forming fluid. According to the hypothesis that the fluid was what was left over from the crystallization of a molten body, the total amount of fluid is limited to the original amount dissolved in the igneous mass and that amount would have had to hold the ore metals in solution. Various geologists suggested different sorts of fluids that might be potent solvents for metal sulfides, with chloride solutions gaining favor. Ore microscopists even found tiny crystals of sodium chloride imprisoned in fluid inclusions (microscopic pockets of fluid) in ore minerals and associated mineral grains. Thus those favoring a juvenile (primary igneous) origin of the ore-forming fluid proclaimed that the solubility problem was a dead issue.

Those favoring some version of lateral secretion insisted that the solubility problem remained unresolved and that other lines of evidence cited by the igneous camp were questionable. The present deposition of ore minerals around fumaroles was a major argument of the igneous camp. In 1912 there had been a volcanic eruption in the Aleutian Islands that filled a valley with a volcanic ash flow. This valley was peppered with fumaroles (vents where hot gases escape from beneath the surface) and was accordingly dubbed the Valley of Ten Thousand Smokes. These fumaroles, which deposited many ore minerals, were cited as evidence by the igneous camp who maintained that the source of the gases and ore minerals was a large igneous body buried under the ash. In the 1950s a team of geologists visited the valley and found that streams had cut all the way through the ash and that the underlying rocks were not an igneous mass at all, but just ordinary sedimentary rocks. The secretionists were elated. The fumarole gases were just rainwater that had percolated into the hot ash, become heated, leached some elements out of the ash, and emerged at the surface as gases that redeposited the dissolved material.

During the 1950s and 60s the use of isotopes for geologic studies became more and more important as techniques became available for the task of isotopic analysis. Using the ratios of stable isotopes of common elements such as oxygen, isotope geologists suggested that by far the majority of exhalations from volcanoes and fumaroles were of meteoric origin (i.e., from rainfall), a concept of long standing that was now given a quantitative basis. During the same period copper deposits associated with volcanic activity were discovered in Indonesia whose ore materials seemed very probably to have been transported by and deposited from seawater that was heated and circulated by the volcanic mass. The secretionists maintained that they had the best of both worlds. Their ore-forming bluid was not only unlimited in quantity because the same water could be circulated over and over again, but wherever marine waters were trapped in sediments, as in Indonesia, the waters would contain chloride as well, explaining the occurrence of chloride in fluid inclusions.

Exploration of the ocean floor in the Gulf of California and north of the Galapagos Islands in 1979 revealed giant hot springs, roaring out of cracks in the floor, that were emitting great black clouds of iron sulfide with dissolved manganese. Immediately around and inside of the vents, sulfides of copper, zinc, and other metals were being deposited. The geochemical properties of the springs indicate that they are caused by heated seawater's circulating through hot, newly formed crust at actively opening cracks in the ocean bottom. The sulfides are deposited in the fissures and also on the bottom around the vents. Ancient deposits of this type have been found on land, such as the copper deposits of Cyprus. Many geologists now believe that large-scale ocean bottom sulfide deposits may be the first step in the concentration of metals into ore bodies. These initial bodies may then be reworked by other circulating solutions, giving rise to the variety of types of ore deposits found in the world. Thus at present geology is dominated by those who espouse leaching while those favoring an igneous origin of the ore-forming fluid are at bay.

What all this indicates is that while theories of origin have become less fanciful, techniques of exploration and mining modified, and innovations introduced in many areas, the theories themselves have not changed fundamentally for hundreds of years and possibly longer. This is less a criticism of modern geologists than praise of ancient prospectors and miners. The definition of what constitutes an ore has changed, however. An ore is not a body of rock that is defined by geologic parameters, as it once was considered. An ore is a body that can be mined at a profit. The definition of ore then depends on the after-expenses value of whatever is being sought rather than how rich it is in metal or other valuable substance. Most mining today is done on a grand scale, so a deposit must first of all be large so that the capital investment in plants and heavy equipment can be amortized over a long period. Most modern deposits are not only large but have low-grade ore.

There are deposits mined today that are not large and lean. I met a bachelor miner in the Sierra Nevada in California who had found an extraordinarily rich deposit of a tungsten mineral called scheelite. His prospector acquaintances had all tried to talk him into getting a government loan to develop the discovery into a major property, but he was wiser than that. Instead, he mined the ore by hand with a geologist's hammer and a 5-gallon bucket. He chipped away at the crumbly rock, filling his bucket, which he then raised from a depth of about fifteen feet to the surface with a hand winch fashioned of wood. This man was a learned, though self-taught, person whose library would have been the envy of any mining geologist. To his credit he had the wisdom to mine with methods that would have been used during the Bronze Age instead of taking the "get big" approach his friends suggested. He didn't even mine until he ran out of cash, at which time he would fill a few buckets and haul them in his Jeep to a nearby mill to sell them. His mine provided him with a substantial income for very little work and no headaches at all.

The value of a given metal depends on a number of factors, but scarcity is certainly one of them, with gold and the platinum metals serving as good examples. Utility is another factor, and it is instructive to discover that in his book on the Red Sea, *Periplus Rubri Maris* (mentioned in Chapter 1), Agatharchides (181–146 B.C.) states that the value of gold on the Arabian shore of the Red Sea was ten times that of silver, three times that of copper, and only half that of iron. This somewhat startling information reflects two factors. First, iron was presumably the most valuable metal because it was needed for swords and knives, and the local Arabs lacked both the ores and the technology to produce their own. Copper is also valued for tool making. Because copper metallurgy is simple, it simply must have been scarce in Arabia. The relatively high value of copper lends credence to the belief of some scholars that Solomon traded copper for gold at Ophir rather than having his men mine it. Copper mines were widespread in Solomon's domain, and there is minor production of copper in Israel even today. The Bible also states in Kings I 10:29 that the cost of imported goods in terms of silver was very high toward the end of Solomon's reign, so silver must have been a metal of relatively low value at the time. When a soothsayer's palm was crossed with silver, it might not have been much of a payment in those times.

The value of an ore may depend on simple recognition of the metals present. Certainly the early miners would have recognized no value in ores of aluminum, titanium, or tungsten because there was no technology to extract them from their constituent minerals. Other valuable materials were simply overlooked, and many old mine dumps and tailings from extractive plants have been reprocessed years later to recover metals left by either ignorance or primitive technology. A really remarkable example of overlooking a valuable material occurred in the early days of mining in the area around Virginia City, Nevada. Miners seeking gold were plagued by a heavy, bluish-gray material that plugged up their extractive apparatus. It was some time before someone discovered that the "nuisance" was silver-bearing minerals. Thus the great silver bonanza of the Comstock Lode was found quite accidentally after being totally overlooked by numerous prospectors. One might say that silver is not only *where* you find it, but *when* it is recognized.

The minerals industry at present increases ore reserves mostly by improving extractive techniques rather than locating new deposits. Through history and even in present times, lower and lower grades are considered ores as the processes of mining and extraction become more efficient. Roughly speaking, if an extractive process improves or the price of a metal goes up so that an ore half as rich as before can be mined, then the reserves are doubled. Of course this cannot go on forever.

Today, any new deposits that are likely to be found will be in polar regions or hidden by jungles, water, or deep sedimentary cover. Exploration in such regions will be successful only if theory is able to measure up to the ultimate and only meaningful test—prediction. So far there is little promise of immediate success. A number of years ago the Canadian geophysicist-generalist J. Tuzo Wilson proposed a grandiose theory relating the location of ore deposits to the major structural blocks of the earth. His theory was expounded before the recent discovery of mineral-laden springs belching out of the seafloor, but that only strengthens its general thrust. A refined version of Wilson's large-scale theory of ore deposition could provide a framework for locating target regions for more detailed exploration. But the targeting will have to be fairly accurate if it is to be useful. The areal exposure of many ore deposits is very small, a matter of a few hundreds or thousands of square meters. In a covered area the only definitive exploration tool is a drill hole, and drill holes are expensive. Economics is the final factor, and a modern exploration program will be unable to survive on accidental discoveries. Gold is still where you find it, as it always has been, and we shall see if twentieth- or twenty-first-century geology is up to the challenge.

What if geology fails? Then, perhaps, we will see history repeat itself. Yale's Brian J. Skinner has suggested that we may enter a new Iron Age as nonferrous metals become increasingly scarce and valuable. Even a common material like copper will eventually have to be allocated primarily for uses that are unique to its properties rather than for electric wire, cooking pots, and the like. Metallurgists will have to develop new iron alloys with properties that will allow them to take over the tasks of other metals. We may even regress further in history and return to the Stone Age by using what amounts to high-technology pottery—ceramic materials— for many purposes that now require metals. Transmission-of-information technology is already phasing out the use of metal wires in favor of glass fibers, and many machine parts are currently made of highly stable and long-lived ceramics instead of metal. It is reasonable to expect that ceramists and other technologists will develop suitable materials for many purposes—but people in lab coats and hard hats sure aren't as romantic as those old prospectors.

Chapter 12

Wind, Sand, and Stuck

S EDIMENTS AND SEDIMENTARY ROCKS FORM A THIN LAYER over much of the crust of the earth. To put the thickness (or thinness) of the sedimentary veneer in perspective, imagine that you have drawn a circle representing the earth that takes up most of a standard sheet of typing paper. Imagine also that this representation is drawn to scale. If you used an ordinary pen or pencil and didn't even press very hard, the width of the line you drew would be about two hundred times thicker than the sedimentary cover over the earth. Of this thin film of sediments, only about 12 percent is made of sand-sized particles, and only a minor fraction of that is the sort of pure, essentially monomineralic sand that we associate with beaches or desert sand dunes.

For more information, see references 4, 44, 53, 64, 74, 79, and 98 in *Further Reading*.

or even attractive to people. Tourist hotels fronting on the whitest beaches command the highest room rents, and Rudolph Valentino wouldn't have been half as interesting to movie fans in the 1920s if he had not been able to grab the heroine and thunder off across the trackless sand dunes on his trusty horse.

In Hollywood's view of the natural world, sand has its dangerous side as well. Many a grade B movie keeps us on the edge of our seats with scenes of someone sinking slowly and inexorably to his death in quicksand in a spooky swamp. Or in the deserts we see an unwary traveler whose vehicle becomes hopelessly mired in soft sand, and who then wanders off, staggering, slipping, and stumbling, in a futile search for water. Unless a Cajun passes by in the swamp or a camel-mounted Bedouin or crusty old prospector encounters the desert wanderer, the fate of strangers to the ways of sand is sealed—at least in old movies.

To people who associate closely with sand, be they geologists or sunbathers, part of its fascination is the very disparate mechanical properties that it can exhibit. Some sand will support vehicles so well that beaches are used for auto racetracks. In other places a car or truck promptly bogs down. Both wet and dry, sand can be firm or a seemingly bottomless trap; sandy regions can thus be insurmountable barriers to travel for the novice, or easy highways for the knowledgeable.

In the early decades of this century, when British influence was still strong in North Africa, a British desert explorer named R. A. Bagnold became interested in the windblown sands of the Libyan desert. He was familiar with the literature concerning desert landforms, but he concluded correctly that the authors didn't really know much about their subject. Bagnold, a graduate of the Royal Military Academy and a geologist with a quantitative bent, wanted some hard facts and numbers, not just the broad generalities and qualitative guesses—termed "arm waving" by disparagers— offered by physiographers, the physical geographers who study landforms. He collected field data for many seasons, getting his car stuck numerous times in the process. He observed that in many places the tire tracks of his car were barely noticeable, whereas in others that looked just the same to him, the car would abruptly sink to the running boards. He found that in these soft spots a 6-foot-long rod could be thrust in effortlessly and that if one jumped in the middle of such a patch of sand, a circular wave would propagate out as in a fluid.

After his field studies Bagnold designed some laboratory experiments to further his studies of sand and sand transport. He built small wind tunnels to study how wind moves sand grains. Prior to his experiments people had taken it for granted that sand grains rolled along or were held in suspension by the wind for long distances. The old guard of desert experts ridiculed Bagnold for trying to duplicate the vast Libyan desert in his little wind tunnels, but he gathered crucial information, secure in his search for real data rather than speculation.

His results were published in 1941 in the classic book *The Physics of Blown Sands and Desert Dunes*, considered by many to be the finest treatise on sedimentation ever written. The fact that new printings of the original text are still being assigned to geology graduate students for reading is testimony to its excellence and permanence. Bagnold proved that sand grains in the desert are transported not by rolling or by being carried in suspension but by a process of hopping called saltation after the Latin for dancing or leaping. In saltation, a sand grain that is lifted into the air by some process is carried a short distance downwind before falling back to the surface. By virtue of the energy it picked up from the wind it strikes the surface with enough force to dislodge one or more new grains, which then repeat the process. The stronger the wind, the more energy is imparted to the saltating grains, so the harder they strike the surface. Thus the stronger the wind the higher the hops. The total effect looks like a cloud of suspended sand but in reality is more like a crowd of frogs all hopping in the same direction.

Dune sands are very well sorted regarding size; that is, they contain only sand-sized particles, not gravel or fine silt.

This is because large particles are too heavy to saltate, and the finer ones get carried away in suspension. Bagnold had put the transport of sand and the form of dunes on a quantitative and therefore firmer and more respectable basis, but he still hadn't answered the question of why his car got stuck in some places and not in others.

Bagnold had observed that in the desert the location of a soft "quicksand" spot in a given dune was seemingly random. However, he noted that in a group of dunes the soft areas were all in the same relative position on each dune.

Dunes have a variety of shapes but generally consist of two parts. First, there is an upwind portion on which sand moves by saltation. This upwind part is commonly streaked or rippled in response to the wind's action. Second, there is usually also a downwind or lee side that has a steeper slope and lacks the surface features seen on the upwind slope. Bagnold and others knew that a dune moved by the grains' saltating up the shallow, upwind side and then rolling down the lee side. In a cross-section of a dune these two regions looked different, with the upwind side showing regular internal layering and the lee side showing either no layering or very crude layering. Such layers were simple enough to observe in dunes that had grains of different size or color, but the trenches Bagnold cut through the Libyan dunes revealed neither. After some experimentation he found that if he poured water over a cut dune surface and then scraped away the wet surface layer, he could see the layering in the dune. What happened was that the water would soak in a little more in some layers than in others because of slight differences in grain size. Those damp layers held together whereas the intermediate dry layers fell away, leaving layering visible to Bagnold's searching eye.

With his water method Bagnold discovered that the soft spots of the dune were lee-type sand masses whereas the firm parts were the layered upwind sand. The soft places were simply old lee deposits that had been buried under upwind layers and that had subsequently become exposed at the surface by removal of the overlying sand cover.

Bagnold attributed the difference in strength of the two

kinds of sand to how solidly the sand grains were packed together. The layered, upwind-type sands were tightly packed together by the hammering action of saltation. Tightly packed sands have less air space between sand grains than unpacked ones, and so the grains have more points of contact with one another. With many points of contact a strong framework results because of the summation of the frictional resistance to movement, the force between a grain and its neighbors that must be overcome before grains can rub past one another. In contrast, the lee-side sand was loosely packed. It had not been subjected to the hammering action of saltating grains but accumulated by the rolling of single grains or the avalanching of masses of grains. The resulting sand mass contains lots of air spaces and much less grain-to-grain contact, so the total frictional resistance is much smaller.

Bagnold now understood why the soft spots were always located in the same relative position in each of a group of dunes. Dunes in a field are all similar in shape, and as long as the dune is not migrating, the upwind and lee surfaces are separated and clearly defined. If the dunes migrate or if there is a seasonal shift in wind direction, then erosion of the upwind surface results in the exposure of the soft, inner lee deposits. All the dunes in a field would be affected in the same way, so the soft spots would have to be in the same relative places. The mystery of the desert quicksands was solved.

It goes without saying that the desert is not the only place where a vehicle can get stuck in the sand. Wet sand can also offer little support to the traveler. The properties of wet sand generally depend on the degree of wetness. If you were to start with well-packed dry sand and add water a little bit at a time, measuring the strength of the sand body at intervals, you would discover that its strength would increase with increasing wetness up to a point, then decrease again.

The first effect—strength increasing with wetness— results from the fact that films of water around the grains hold them together by the molecular attraction of water for the grains and for itself. Anyone who has walked on a wet beach knows that moist sand is strong, making a fine place for walking or running. Beach sand is tightly packed by the pounding action of the surf, and the force of a walker's foot shifts the grains a little from their tightly packed arrangement. As a result the pore space between the grains is increased, causing the water to retreat from that area and hence forming the visible "drying up" of the patch of sand just under your foot as you walk.

If the sand contains more water than is just enough to coat the grains and add strength, then the effect of buoyancy begins to weaken the sand. As the philosopher Archimedes was said to have discovered in his bathtub, an object is reduced in weight by an amount equal to the weight of the immersing fluid displaced. Sand grains have a density about two and a half times that of water, so the result of their immersion in water is a weight loss of about 40 percent. Since the grains weigh less, they press less firmly against one another and the friction between them is reduced. The water also acts as a lubricant to reduce friction. The result is a thoroughly wet mass that has low strength. In addition, relative to a dry body of sand, soil, or unconsolidated rock, the wet mass weighs more. The combination of lower strength and higher weight has caused many a mountainside to slip away during heavy rainfalls, especially in places like California, which abounds in suitable, unconsolidated rock bodies that are exposed in steep slopes.

In some saturated sands—sand bodies in which *all* the pore spaces are occupied by water—water may move through the sand mass. If that movement is upward, then the stage is set for the man-trapping quicksand of the movies. Saturated sand exhibits the buoyancy mentioned above plus friction between the moving water and the sand grains. If the upward water movement is fast enough to lift the grains in suspension just to the point of moving, but not quite, the sand is said to be "quick." If the water motion is faster, the sand actually moves and the sand–water mixture "boils." Sand–water mixtures that are in motion are likely to be noticed by an observant passerby, but sand that is quick, but not quite in motion, appears to be a perfectly solid mass of sand—yet it has absolutely no strength for all practical purposes. A rock thrown into quicksand will vanish from sight as if thrown into plain water, and a person stepping onto it will drop abruptly to a level somewhere between the waist and armpits, with no solid footing underneath, just the sensation of a bottomless pit.

The reason the rock sinks, whereas the human stays afloat, is explained by the relative densities of the two substances. The density (weight per volume) of the quicksand is a combination of that of water (one gram per cubic centimeter) and that of the sand (about two and a half grams per cubic centimeter), its exact value depending on the proportion of sand and water. The rock has a density of about two and a half, and sinks out of sight. A human has a density of less than one and will bob like a cork in the heavy sand–water fluid, as I discovered when I plunged unexpectedly into an apparently solid sandbar in a Sierra Nevada creek.

After I overcame the initial astonishment of standing on nothing and floating about in the chilly, watery sand, I threw my hammer and map case to the shore and swam and scuttled to solid sand nearby where I could stand again. Looking back at the place where I had been floating I could see no indication that it was anything but a solid sand bar.

If I was able to get out so readily, then why is the man in the movie always slowly sucked down to his death? The answer is that I grasped the physics of the situation and got out promptly. The person who sinks does so because his density gradually increases until it exceeds that of the quicksand. How his density increases is indicated by the fact that I found every nook and cranny of my clothing full of sand—pockets, the seams of my clothes, my boots and socks, sand everywhere. Fortunately I was in a remote spot so that I could strip to the buff and wash my clothes in a clear part of the stream and hang them on a bush to dry, for walking any distance in sandpaper clothing would have been anything but pleasant.

The sand gets into your clothing because the clothing slows the motion of the water. Consequently the sand is no longer held in a weightless state and sediments into pockets, shorts, and whatnot. Thus if a person stays in the quicksand very long, struggling to find solid footing instead of trying to swim out, the accumulation of sand will pull him down and under.

Saturated sand bodies with water moving through them can exhibit physical properties that even Hollywood would hesitate to show to its viewers for fear that nobody would believe them. Sand that is loosely packed like Bagnold's soft spots and also saturated is highly unstable, even more so if the water is in motion. The friction between grains is negligible, and any sudden stress can cause the mass to behave like a fluid. The remarkable transition from solid to fluid occurs when the mass is abruptly disturbed by pile driving, explosion, or earthquake. As the shock waves passes through the mass, individual sand grains move and then sink to new positions. Any single grain may only shift by a tiny amount, but during the time the grains are in motion there is a momentary total loss of cohesiveness and the mass behaves like a liquid of high density. That is, for a short time it acts like quicksand.

During the brief interval that the mass becomes liquid objects behave as dictated by their density. Large boulders and manmade works such as statuary and buildings can sink out of sight, vanishing without a trace as if swallowed up by the earth. There are reports of objects vanishing in Japan. Materials that are less dense than the fluid float upward. During one earthquake in Japan the pilings of a centuries-old bridge rose abruptly out of the ground as they floated upward, leaving the bridge on high, unsteady stilts. In Chile, underground gasoline storage tanks suddenly became surface gasoline tanks when an earth tremor made the burial site temporarily liquid. During some nineteenth-century earthquakes in India and also in the great Madrid, Missouri, quake of 1811, sand–water liquids squirted out of the ground to build little sand volcanoes.

Engineers are cautious about siting dams over permeable rocks or soils because when the dam fills, water may migrate under the dam through the underlying rocks. If it consists of unconsolidated materials, the footing base that was solid when dry or just moist may become quick when water saturates it and flows through it.

The phenomenon of quickness is not restricted to rocks with sand-sized particles. In parts of Scandinavia and near Ottawa, Canada, there are rock masses known as quick clays that can be mobile. These bodies are made of very fine-grained particles. They can be firm and solid like a potter's clay but can also become syrupy liquids when affected by shock, or sometimes seemingly by caprice. The clay's behavior is somewhat like cornstarch in water. If left to settle, cornstarch forms a very solid layer in the bottom of a measuring cup. If stirred, though, it makes a thick liquid that pours readily.

Quick clays move when shocked, though even a small slide in one part of a mass can start a disturbance that will set a much larger mass of earth in motion. It is said that Norwegian farmers are very watchful of the small slides that they encounter along riverbanks or in ravines because they know that little ones can become big ones and carry their farm—house, barns, and all—down the river. I visited a small quick clay slide in Norway, and it was quite a sight. What had been pastures was now a vast amphitheater about 150 meters (500 feet) across with walls up to 8 meters (25 feet) high, floored by dark, bluish-gray clay. The surface of the clay floor was smooth, as if burnished by the quick clay as it left. Part of the lost mass of the farm was scattered along the banks of the adjacent river, but most of it had been swept downstream.

The slide I visited was a small one by Scandinavian standards. In 1978 a farmer near Rissa, Norway, dumped a load of excavated earth onto the edge of a lake. The impact caused the subsurface clays to start liquefying and within a few minutes 81 acres of farmland and a number of buildings slid and flowed into the lake. In 1893 a quick clay slide in Verdal, Norway, removed or destroyed over 3 square miles of a settled area, killing 120 people. Another in 1950 in Surte, Sweden, wrecked the homes of three hundred people and moved a highway and railroad together with over a million cubic feet of soil into the nearby Göta River. Closer to home, a slide about the size of the one I saw in Norway moved a school and several other buildings into a river at Nicolet, Quebec, in 1955.

Quick clays are structured differently from quicksands, but the lack of tight packing of the particles is common to both. Clay minerals are tiny flakelike crystals. In a potter's clay, the plasticity can be attributed to the flakes' sliding against one another. In quick clays the flakes are arranged like a house of cards, with lots of spaces and hence very little friction between the flakes. Such arrangements of clay particles can be relatively stable if the voids are filled with water containing dissolved salts. The electrical attraction of the ions in the pore water to the clay particles serves to bind the particles together. The quick clays of Scandinavia and eastern North America were deposited under ocean waters during the recent glacial ages. After the glaciers retreated the land rose, placing the clays above sea level. Over the thousands of years since that time, percolating rainwater has removed the soluble ions in the pore waters that once bound the clay particles together so that they are now quick clays.

I suppose the thought would have been little comfort to Bagnold when his car was thoroughly stuck in a sand dune for the umpteenth time that season, but at least he wasn't watching his farm change into a gooey, black liquid that carried buildings and livestock down a nearby river. The next time someone uses the cliché "solid as the ground we're standing on" just let them know that you have a couple of stories to tell them. And as a rule of thumb, don't step on sandbars until you test them, and for goodness' sake don't invest in streamside real estate around the St. Lawrence, Ottawa, or Saguenay River valleys without taking along a geologist to examine the site, preferably a Norwegian or Swedish one. Better yet, take an old Norwegian farmer.

Chapter *13*

Scenario
in the Sediment

*I*N CHAPTER 9 I GAVE YOU SOME IDEA OF THE APPROACH A PALEON-
tologist might take to the question of how many individuals of an animal of a given size might have lived at some
time in the past. Such a determination is an estimate at best,
but it can be tied to a fairly solid anchor point, such as the
amount of incident solar radiation, which acts as a limit on
the total possible biomass. In the case of Loch Ness monsters
or Nebraska "elephants," an estimate of total biomass for
their size range gives the biomass of a single kind of animal
because the unusually large size of monsters and elephants
makes them the only animals in that size interval. For most
animals this is not the case, with the possible exception of the
tiny shrews, the only animals at their end of the mammal size
range. For the great majority of species the paleontologist's

For more information, see references 8, 9, 40, 49, and 91–94 in *Further
Reading.*

148

goal is to use the fossils and their enclosing sediment to put together a picture of the total life environment, including the proportion of each of the now-extinct species to the total population that may be presumed to have coexisted in some sort of loose community. This endeavor can be called *paleoecology*, if you like.

There is also a discipline called *taphonomy*—meaning literally "the laws of burial"—to which we must first turn if we are to reconstruct past communities and environments. Let me state at the outset that taphonomy is not a branch of paleontology. Perhaps the best way to characterize the relationship between the two areas of study is to state that very few paleontologists are taphonomists but that most taphonomists are paleontologists or biologists with a paleontological bent.

Taphonomy is to paleontology what bouillabaise is to plain broth in that it makes use of many and diverse bits of information to create a total picture of the environments of birth, life, death, and burial of extinct organisms whereas paleontology in its simplest form is little more than the description and classification of fossils. In 1976 a group of eighteen scientists met together at Burg Wartenstein in Austria for a symposium sponsored by the Wenner-Gren Foundation to discuss and develop solutions to the difficulties encountered in the practice of taphonomy. While it is difficult to assign each participant unambiguously to a discrete discipline, from my perspective as a geochemist the group consisted of six paleontologists, five biologists, three archaeologists, two geochemists, a geologist, and a geophysicist. That diverse group of scholars attacked the concept of taphonomy during ten days, averaging about sixteen hours of work each day, and we came up with a lot of ideas that resulted in a book called *Fossils in the Making*. I think it is fair to say that we established no set of rules for the practice of taphonomy except that you must use every possible bit of evidence available if you are to come even *close* to characterizing an ancient burial from the fragments and scraps that are usually available.

Perhaps the reason we came up with no manual for begin-ning taphonomists is that every case is unique in one or more ways. The taphonomy is fairly obvious at some sites, includ-ing the fossilized stands of trees that have been found in a number of parts of the world, the insects preserved in the natural flypaper of tree pitch changed to amber, and Pompeii and Herculaneum, whose residents were killed and buried under volcanic debris in the midst of their ordinary daily life, surrounded by the objects of their time.

Most fossil occurrences are not so simple to interpret, and it is no surprise that many early paleontologists made inter-pretations that were later found to be incorrect. In all fair-ness, though, the pioneer fossil hunters were generalist naturalists and did at least attempt to make reasonable interpretations of what they found. As the old-time natural-ists passed on there was a period in the second quarter of the present century when paleontologists became specialized, spending most of their energy collecting and describing and nothing else. They were little more than collectors of objects whose main concern was to increase their collections and to name and classify previously undescribed creatures.

In the early days of American paleontology there was great interest in the natural history of the unexplored western part of the country. Thomas Jefferson, a man who would have fit in well with Scottish geologist James Hutton and his circle of Edinburgh intellectuals, sent Clark of Lewis and Clark fame to Big Bone Lick in Kentucky to bring fossils, especially those of mammoths, east for study. Later, in the nineteenth cen-tury, the paleontologists O. C. Marsh and E. D. Cope scoured the plains and Rockies in a famous competition to see who could describe more new species as well as enrich their home museums at Yale and Philadelphia respectively. Even in *this* century the abundance of vertebrate fossils in the west was nothing short of staggering. My late colleague Paul McGrew described to me his first visit to a locality in eastern Wyoming where I had collected samples for a geochemical study. On his initial visit in the 1940s, he was able to stand on a low rise in the midst of the badlands clays and from that spot see a full dozen complete titanothere skulls exposed at the surface. Such surface exposures are rare today because of extensive collecting by both scientists and hobbyists. Still, at some sites in Wyoming and Nebraska that I have visited you can collect dozens or even hundreds of surface fossils in a day, although the specimens are far smaller than titanothere skulls, which weigh 50 pounds or more.

Collecting in virgin territory, as Cope, Marsh, McGrew, and others did, must have been very exciting and great fun besides; but just collecting fossils to describe, name, and possibly display them is a long way from either taphonomy or paleoecology. Modern scientists sometimes speak of the terrible waste that characterized the old-time collecting methods; the fossils were commonly collected, crated, and sent home without much knowledge of the geologic context of the specimens. The stratigraphy, or regional arrangement of the layered rocks that contained the fossils, was imperfectly known, and generally the fossils were simply labeled as having been collected from a given geographic location without regard to stratigraphic level or any other details. The scientists of the time were too busy with the discovery of the wondrous dinosaurs of Mesozoic time and the spectacular array of mammals that followed them during the Cenozoic to be concerned with much beyond description of their finds. While it is doubtless true that much more information could have been gathered along with the fossils, most of the early finds were already exposed at the surface so that much of the geologic context had been washed or blown away by natural processes before the paleontologists arrived at the site. Consequently surface fossils are pretty useless for more than just descriptive biology, so we need not rue their loss to science.

A fossil deposit is a naturally occurring sample of an ancient population, so the taphonomist or paleoecologist must make use of statistical methods. Statistics was developed and is still used mostly for making mathematical statements about large populations from small samples. For any statement about a population to have validity in the eyes of a statistician the method of sampling must have followed some well-defined rules, the most important of which is that the sample must not be biased in any way but, at least ideally, must represent each element of the population in proportion to the ratios in the total population. The sample must be a so-called random sample. The famous incorrect prediction of pollsters in 1948 that Thomas Dewey would defeat Harry Truman for the U.S. presidency was based on a nonrandom sample of voters because the survey was made by telephone, which biased the result against the votes that might be expected from persons who didn't have telephones, a fairly large group in those days. Similarly, samples of a factory product collected at the end of a shift would be nonrandom and hence unlikely to be representative of the products made over the entire shift.

Nature is unconcerned with the desires of statisticians or paleontologists, and we have every reason to believe that most natural samples are biased in one or more ways. For instance, although the mode of burial is clear, we cannot be sure that the sample of Italian life preserved at Pompeii is representative of Pompeii, much less of life in Italy at that time. Perhaps when volcanic ash overwhelmed the town one or more groups were absent. Perhaps the swells were off to the beach for a party, and maybe some of the servants had gone to town to buy food or other household items.

Vertebrate fossil sites are biased in many ways, the most obvious of which is that the burial site cannot possibly preserve fossils from a large region in any uniform way. You need only think about modern environments to confirm that statement. If the deposit is one made by a slow-flowing river in a broad valley, then it is unlikely to contain specimens of animals that lived in nearby mountains or of those that required little water and stayed on the plains. You could argue that river deposits are a reasonably good sample of a large area because during heavy rains many bones from a large area would be washed into streams and thence to the river, thereby constituting a broad sample. There is some truth in this, but there are major problems too. For instance, transportation by water includes sorting by hydraulic properties. Some bones move more easily than others because of their size or shape, as has been demonstrated by Mike Voorhies in a laboratory flume and by Kay Behrensmeyer in a Wyoming river. Also, a great flood that would sample a large area is also likely to excavate bones that were buried in some earlier depositional event (a flood, etc.) and mix them with those from living species.

Another type of site is the spring site where animals either become trapped or die of other causes when visiting the spring for water. That sort of site is presumably more representative of animals that drank together at the waterhole but excludes species that obtained their water from food or from dew or other sources. This is the same kind of nonrandom sample you would have if you studied patrons trapped in a bar or soda fountain in Pompeii rather than studying the whole town. In an even more selective preservation, Bob Hunt of the University of Nebraska found specimens of bear dogs that were trapped and fossilized in their dens near Agate, Nebraska, much as if a single family of Pompeiians had been caught in a hotel or condominium.

Deposits of vertebrate fossils commonly show striking departures from what you might expect at first. One of the famous localities near Agate, Nebraska, contains virtually nothing except remains of three species: a rhinoceros (90%), *Moropus* (9%), and *Dinohyus* (1%) (sorry, but there are no modern analogues to the mammals *Moropus* and *Dinohyus*). Our gut reaction is that surely there were other mammals living in that area when the fossils were buried. Another locality northeast of Agate is even more peculiar, containing solely fossils of a relative of camels called *Stenomylus*. Still another quarry near Verdigre, Nebraska, contains almost nothing but rhinos. Another near Torrington, Wyoming, yields essentially nothing but remains of two species, a rhino and a horse. The list of sites could go on almost indefinitely, but the point is that it is common to find fossil assemblages that are composed of only one species or very much dominated by one or a few species.

Such mass occurrences must represent mass deaths of

large flocks of animals, so taphonomists have concluded that a flocking tendency must be characteristic of the species found in these settings. In addition to that behavioral trait, we must assume that the flock was all killed at much the same time as well, especially if the fossils show little weathering, teeth marks of other animals, and/or indications of extensive transportation. Andrew Hill of the Leakey Institute in Kenya told our group in Austria about witnessing just such an event in east Africa in which a large herd of wildebeest attempted to cross a river in flood and were mostly drowned and buried in short order.

Mass burials, then, are of little use in reconstructing the overall makeup of an ancient community of animals except to the extent that they indicate which species were flockers rather than loners. The mass occurrences are most useful in studying the properties of an individual species, such as age distribution within the flock, since it is probably fair to assume that if the whole flock was not buried, at least a representative (random) sample was preserved.

An interesting feature of the mass burial phenomenon is that a species that is very abundant at one site may be absent at another. The pronghorn antelope–like *Merycodus*, of which Mike Voorhies found over 2,200 specimens representing over 500 individuals at his Verdigre NE site, was totally absent from a collection made by American Museum of Natural History paleontologist Morris Skinner at a site in the same rock unit about one hundred miles west that is nearly identical in age and contains many species also found at Verdigre NE. We must assume that a *Merycodus* herd simply was not in the vicinity of Skinner's collecting site at the time of burial of the fossils.

On the other hand, there are fossil species that tend to occur in small to moderate numbers in many deposits. In the Cenozoic rocks of Wyoming and Nebraska one of the more widespread fossil mammals found is a piglike animal called an oreodont, which is present in small numbers in a very large fraction of fossil occurrences. Because of this pattern it is assumed that oreodonts must have lived in small family groups rather than large flocks. The oreodont fossils are commonly associated with the remains of rodents, rabbits, turtles, and a few carnivores, forming what might actually be a fair representation of death by normal attrition of the common animals that inhabited the region. The University of Wyoming collections include a group of oreodonts, huddled closely together and buried in place. Such deposits are considerably less spectacular than those of mass burial origin but can be more useful to the paleoecologist.

Still, just listing species present does not give much of a clue to the living populations that existed in the past. If it did, the collections made in the last century would be adequate, which they are not. As Mike Voorhies so aptly puts it, "Taphonomic studies should be started at the time the collection is made. Once a fossil is out of the rock at least half of the critical clues are gone." What sort of clues are there to find?

The enclosing rock of a fossil is called the matrix and is normally a sediment that may be more or less hardened into rock. The nature of the sediment tells the taphonomist quite a bit. A fine-grained, clayey sediment means that the water motion at the site of deposition must have been very slow to nonexistent because tiny sedimentary particles are easily washed away. It might also indicate that the grains were deposited during a dust storm. A sandy matrix means that moving water winnowed out the very fine fraction and would also have been able to transport some of the fossil material. A gravelly matrix means that very actively moving water was able to remove much or all of the sand-sized particles as well as the still finer grains. The nature of the layering in the rocks is also an important clue. A few thick layers might reflect rapid deposition of a lot of sediment during brief episodes whereas fine, uniform layering or lack of layering might result from a steady, slow accumulation.

Another feature that can be observed only while the fossils are still in the rock matrix is the mutual relations among the bones and other skeletal elements. If an animal died and was buried in situ, at the place where it is now found, then parts of the skeleton should be found in close association. In some localities, skeletons are found with the skeletal parts still in the life arrangement, rather than disarticulated, or disjointed. Such finds, which as one might expect are uncommon, leave little doubt that the animal was buried immediately after death or even killed by the burial.

More commonly, skeletons are found with most of the parts closely associated, but with some scattering and some elements missing. Such fossils are interpreted as having been buried more or less at the site of death, but after a period of exposure to the atmosphere. People who have studied the disarticulation of skeletons find that the process is quite rapid with small- to medium-sized animals. It has been my experience that when a sheep dies in one of our pastures it disintegrates so rapidly that it becomes lost in the grass in a matter of days, and the shepherd searching for it has to almost walk over it to spot the corpse. Andrew Hill of Kenya's Leakey Institute observed that in Africa the flesh is gone from medium-sized animals in a day or two. Malcolm Coe from Oxford studied the disintegration of elephants and found that even an animal that large is fairly decomposed and flattened after three weeks, strictly bones after a year, and partially buried after two years. In all cases, many of the smaller or less massive bones vanish, presumably eaten by scavengers of various sorts. The remaining bones are commonly grooved by teeth marks and show effects of weathering. The rate of disintegration is highly variable, depending on factors such as humidity and temperature, with dry climates favoring slow breakdown. At our Minnesota farm the rate of disintegration of a dead sheep is very much dependent on the season of the year. In spring and fall, migrating crows in the area strip the carcasses to bare bones in a day or two, and at the peak of summer, consumption by fly maggots does the job in a matter of about three days.

Fossils that are preserved at the death site are the exception, for most are moved to some degree by natural processes, mainly the movement of water. Water motion may winnow away the more easily moved bones, leaving the less tractable ones behind. Such a process tends to twist and turn the bones that remain so that they become aligned with the water motion. The absence of small bones together with the alignment of the remaining ones is a sure indication of the work of moving water.

Most vertebrate fossils are transported to some degree prior to burial, resulting in the separation of skeletal parts, the mixing of bones from different species and individuals, and the like. The corners of fossils that are transported very far are abraded and rounded, and the degree of abrasion is a rough guide to the amount of movement. It is common to find fossils associated together that show very disparate evidence of transport.

Typical transported deposits contain a middle range of bone sizes, the largest having been left behind and the smallest swept onward downstream. This is the case in the beautiful taphonomic study that Mike Voorhies made at the Verdigre NE site in Nebraska. Large elements such as rhino skulls or mastodon pelvises are absent, as are all portions of the skeletons of small animals like rodents and insectivores except for their lower jaws, which are less easily moved by water. This sorting effect of water transport is confirmed by the fact that more types of skeletal elements have been recovered for animals in the middle size range. As Voorhies has demonstrated by placing bones in a flume that simulated a stream and observing their behavior, all these elements were about equally transportable.

Transported deposits have another feature that sets them apart from ones formed in situ—the occurrence together of species that would seem to be unlikely associates from other evidence. One example of this phenomenon is seen at Verdigre, where the fossils of two horses are represented about equally even though from the character of their legs and teeth we can conclude that one of them—*Protohippus*—doubtless lived life in the fast lane, racing about the open spaces and subsisting largely on grasses, whereas the other—*Hypohippus*—was a stocky, slow animal who undoubtedly browsed on leaves and lived in wooded areas or swamps. How can their coincidence be explained? One possibility is that bones of *Protohippus* might have washed into the streams and swamps to be preserved along with the resident *Hypohippus*. In this case we would still expect the *Hypohippus* fossils to outnumber their speedier cousins. A more reasonable explanation is that *Protohippus* had to visit the home turf of *Hypohippus* at least once a day to get a drink, and some individuals died while near the stream.

Horses, rhinos, and their relatives are dependent on regular access to a water supply because they have limited storage capacity for it and also do not retain it very effectively, especially when compared with ruminating animals. Ruminants can not only retain a lot of water in their various stomachs but even manufacture some water in their rumen. In addition, ruminants are able to recycle urea and uric acid into their rumen for use as a nitrogen source whereas the single-stomached horses and rhinos must excrete it in urine, thereby expending some of their water supply.

Thus at a site like Voorhies' Verdigre NE quarry we find an association in death of residents of the river bottoms and swamps like *Hypohippus* together with thirsty forms like *Protohippus* who actually spent most of their lives in grassy areas. Doubtless *Protohippus* shared that domain with the little *Merycodus*, which was probably a water retainer and should not be especially well represented in the fossil assemblage. Yet *Merycodus* specimens account for roughly 80 percent of the fossils. The explanation of this disparity is that flocking animals tend to be represented massively when they are represented at all; it must be that a flock just happened to be at the river for some reason when a flood occurred. You might say that flocking animals are disaster-prone.

A realist would have to admit that such deposits make it pretty difficult to reconstruct a past environment even though we can at least get some idea of the mammalian community at the time. Birds are rarely preserved at all because of their small size and fragile bones, as is true of most small reptiles and amphibians. There are probably also large mammals that are rarely if ever preserved because of peculiarities of their metabolic makeup and consequent life-style. A modern African mammal that would seem to have little prospect of being preserved is the delicate, lovely Grant's gazelle. For the most part, this gazelle lives alone or associates in small groups, so it is not likely to be buried in a mass grave. In addition it is water independent to an amazing degree. Not only does it not need to drink much water for digestion and the like, it does not even need water for cooling because it has a set of interconnected chambers in its skull that allow it to pump air past tissues that keep its blood and brain cool. Thus Grant's gazelles graze in midday on the searingly hot flats of the Rift Valley in east Africa where no other animal could survive. Their special niche means that they have a very small likelihood of being buried in river sediments for potential fossilization. There must have been similar mammals in the past of which we understandably have no fossil record or that appear to be rarer than they really were in life.

In any collection of fossils of the same species the skeletal elements come from animals of different ages. We can judge age in a rough way from the size of the bones, although sex strongly influences bone size in many species. The most useful skeletal element for estimating age is the skull or the more commonly found lower jaw. The distribution of individuals by age is a useful tool for the paleontologist, and like any tool it can be used correctly or incorrectly.

Vertebrate paleontologist John Clark wrote a paper in the late 1930s in which he described what he termed a crisis in the evolution of horses. Clark had examined a lot of jaws of *Mesohippus* and noted that most of them showed very little wear of the teeth. He concluded, rightly, that the wear distribution meant that many juveniles were perishing but then went on to conclude that this line of horses was in danger of becoming extinct during that critical span of time. Yet horses made it somehow, and we can conclude either that Clark was right and they squeaked through a difficult time or else that there is another explanation for the high incidence of fossils of juveniles.

Wildlife biologists have studied age distributions among living animals of many types and have found that for most species the most abundant age group represented is the youngest, with a more or less steady decline in numbers with increasing age. These biologists and some paleontologists have also studied patterns of dying among members of animal communities. As I think any of us would expect, they find that there is a high mortality of the very young, then a sudden drop for what we might call youth and early middle age, then a gradual increase in mortality with age. It is worth noting that the age distribution noted by Clark would fit the normal death-by-attrition pattern without giving us any reason to suspect that *Mesohippus* was in danger of extinction. Still, Clark thought that the number of jaws with unworn teeth was unusually high. I think there is a better explanation than Clark's.

Modern horses seldom have twins, a trait they share with domestic cattle. In contrast, sheep and goats regularly have multiple births. These familiar examples illustrate a biological rule of thumb: large animals tend to have low prolificacy compared with smaller ones. *Mesohippus* was a small horse, so it is highly probable that it had multiple births as a matter of course. In addition, modern female horses show no seasonality of their estrus cycles so we might suspect that *Mesohippus* not only birthed lots of twins and triplets but probably had several litters a year. We might say, using the biologist's term for it, that *Mesohippus'* "strategy" was to have lots of offspring that enough survived to continue the species. Since early horses like *Mesohippus* were not fleet of foot like present-day thoroughbreds, they were likely easy prey for the carnivores of the time, so they just substituted reproductive efficiency for speed.

Age distribution data are useful for determining the mode of death of ancient populations. If animals die by normal attrition, then the age pattern should be as described above, with lots of fossils of juveniles, few for animals in their prime, and larger numbers with increasing age. Voorhies studied the age distribution of oreodonts from an eastern Wyoming locality of the sort I described earlier in which other evidence leads us to suspect that the fossils found resulted from normal attrition of a resident community. Sure enough, the age distribution is one of lots of juveniles and lots of "old timers," as Voorhies called them.

At Voorhies' Verdigre NE site the pattern of the flocking *Merycodus* is like that of a living community, with the highest representation of fossils from juveniles and then a gradual, smooth decrease with increasing age. Thus the catastrophic death and burial of a flock of *Merycodus* was a meaningful sample of a whole life community. It is also of interest that the grazing horse *Protohippus* shows the same sort of distribution at Verdigre NE, suggesting strongly that it was a flocking animal as well.

If the information leads to the conclusion that a deposit is of catastrophic origin, then the fossils give, in Voorhies' words, "a snapshot" view, an "instantaneous census" of a particular place at one time, which may or may not be of any use in coming to paleoecological conclusions. In contrast, a deposit of fossils that can be shown to result from attritional mortality is a "time exposure" and as such is more meaningful in paleoecological reconstructions as long as the length of the time exposure was long enough to allow for the mixing of conditions and animal populations.

The last time my wife emptied our laundry basket she discovered a dead mouse in the bottom, beneath the clothes. The discovery was not unusual, for we have periodic invasions of white-footed mice in our country home. The cause of death of the mouse was likely the poison baits that we have around, but my wife and I had a difference of opinion as to the manner in which the mouse got to its burial site beneath the towels, socks, and underwear. I, being a geologist, followed Steno's principles and assumed that the mouse had fallen in and been covered by an accumulation of dirty laundry. My wife took the view that it had crawled in through a crack in the bottom of the woven bamboo hamper to die in the privacy of the linens. Unfortunately, the burial site was disturbed when the laundry was removed for washing so we may never know the answer to this fine example of a taphonomic question, but such is taphonomy.

Chapter 14

Every Moment Is Unique — Almost

*E*VERY STUDENT WHO STUDIES GEOLOGY IS AT SOME TIME SUB-jected to the statement that the present is the key to the past and told that this concept is the basis for modern geologic thought, as based on the teachings of James Hutton and Charles Lyell. Everyone, no doubt, has a different interpretation of this statement. One student may take it to mean that all earth processes are very slow and gradual, especially if she or he lives by a lazy river in the continental interior. Yet another who lives in California or New Zealand might vizualize earth history as a succession of earthquakes, landslides, and other more dramatic events, and a resident of Japan or the Pacific Northwest might view volcanic activity as a necessary part of the past. Each student is correct in some ways, but they may also fall into the error of taking the present as the key to the past a little too literally.

For more information, see references 2, 3, 7, 24, 28, 34, 35, 37, 52, 65, and 73 in *Further Reading.*

It is doubtful that either Hutton or Lyell ever actually used this cliché at all—even though it is supposedly a statement of the principle of uniformitarianism. They and their supporters did think that present-day processes also operated in the past, as the phrase implies, but it is well to remember that uniformitarianism as a philosophical concept was introduced partly to counter the dogma of the church, which held sway at the time and led many scientists to espouse catastrophic change rather than change by small, gradual increments. Abraham Werner of Freiberg in Saxony and his followers insisted that the rocks of the earth had been deposited by worldwide floods of which the latest was Noah's. The French scientists Baron G. Cuvier and A. D. d'Orbingy held that the succession of fossil communities observed in the geologic record represented multiple episodes of catastrophic worldwide extinction and subsequent creation. The notion of the earth's great antiquity had not yet been given much attention.

The uniformitarianist followers of Hutton—most notably John Playfair, whose lucid account of Hutton's concepts in *Illustrations of the Huttonian Theory* made up for Hutton's ineptitude at writing—were really taking a stance against catastrophism and the universal flood concept. Hutton's laboratory was the field, and it appears that he observed with a reasonably open mind and tried to learn from the rocks rather than fit what he saw into a biblical mold. It would seem that Hutton thought that there were cycles of rock formation and destruction that had been repeated many times in the past, whereas his opponents preferred to explain what they saw by unique events—miracles, if you will.

In an argument each side often exaggerates its position, and uniformitarianism, the concept that present-day physical laws and processes operated in the past as they do at present, turned into a more extreme doctrine that might better be called gradualism, a belief that geological change came about by slow steps of almost imperceptible change. Many of Lyell's followers imagined that there were few if any catastrophic events in earth history, whereas the followers of a more nearly biblical version of the past believed in the concept of vast changes brought on by violent forces unlike any observed at present.

Stephen Jay Gould wrote a paper a number of years ago in which he suggested that uniformitarianism was no longer a useful or indeed necessary concept because any rational person realized that the conflict between scientists and Christian fundamentalists was a dead issue. Instead of rejecting uniformitarianism altogether, it would perhaps be more useful to redefine it in light of modern principles. For the present, a useful approach would be an acceptance that the fundamental physical laws that operate today also operated in the same way in the past. Thus the laws of thermodynamics, the speed of light, the scale of time and of rates, and other basic rules and constants do not vary with time. Such a view is equally consistent with uniformitarianism or catastrophism and only demands that every event in history be explained by the ordinary laws of natural science. Such a view is implicit in the radiometric dating of rocks, the interpretation of sedimentary rocks in light of the laws of hydraulics, the use of laboratory experiments to simulate conditions deep in the crust or at past times, the interpretation of fossil communities, and nearly every aspect of modern earth science. In the view of most of today's scientists, such a version of uniformitarianism leaves no room for miracles because every event that has ever occurred has some explanation within the framework of natural laws.

As I have pointed out elsewhere in this volume, there have been events in the past that have happened only once and will very likely never occur again. The initial heating of a cold earth was a one-time event and cannot repeat itself because the conversion of gravity to heat can only take place once and a major part of the supply of radioactive energy is irretrievably gone. In all likelihood the atmosphere will never again be oxygen-free (although it should be noted that there is an oxygen sink in the solid earth in the form of fossil carbon compounds and ferrous iron ample enough to combine with all atmospheric oxygen if the two were allowed to react).

Time is unidirectional, and although the same physical

laws are in operation throughout time, there is never an *exact* return to a past condition. The earth and all its parts have undergone continuous change. Any given event in either the inorganic or organic world is the response of a previously existing microcosm to new conditions, so each resulting life form or rock or geologic feature carries the heritage of all events that preceded the latest one. The unusual gold deposits and iron ores of the Precambrian eras are unique, and others of the same type will not be formed in the future. There will never again be trilobites roaming the seafloor, nor will there be sandstones that look *exactly* like the Cambrian ones. (Present-day sediments on the seafloor are infested with burrowing organisms, as have been all sediments during the last few hundreds of millions of years. Burrowers were of far less importance during most of the Paleozoic Era, and according to the burrower-expert Adolf Seilacher of Tübingen, were altogether absent in all but the very latest part of Precambrian time. Thus sediments of different ages look, and are, different because of differences in the activities of burrowers. We are unlikely to have a disappearance of burrowers in the future.) The presence of land plants has totally changed the course of rock weathering by virtue of the action of their metabolic and decay products as well as the physical action resulting from the growth of their roots and other parts. All this just means that history, of the earthy type at any rate, does not repeat itself, at least not exactly.

The concept of gradualism was probably most strongly adhered to by earth scientists studying sedimentation, in some cases even as late as the 1950s. Their working concept was that particles were deposited one at a time in a slow and orderly way with vast thicknesses of sediments accumulating only after the passage of a suitably vast length of time. These scientists also assumed that all layered sedimentary rocks were originally deposited with horizontal layering when in fact many sediments were deposited in layers that dip, at least moderately, from the outset. They recognized the fact that many sediments might be eroded soon after their deposition but thought it was of little quantitative importance.

Things began to change in the 1950s when the concept of

underwater mudflows was born. Through experimentation in a laboratory flume the Dutch geologist P. H. Kuenen demonstrated that a sediment-water slurry could move as a continuous mass under water and transport a large volume of sediment in a short time. Similar underwater sediment slumps were observed by skin divers off the California coast near the marine station at La Jolla. Bruce Heezen and Maurice Ewing of the Lamont Geological Observatory reviewed some data from 1929, when a number of transatlantic communication cables on the continental shelf off New England were broken soon after an earthquake. They found it interesting that the cables did not all break at the same time but did so in a regular sequence from northwest to southeast. Calculations showed that the sequential breakage was best explained by the movement of an underwater sediment mass of the type produced by Kuenen and observed in California. These sediment masses were called turbidity currents.

It was not very long before sequences of sedimentary rocks all over the world were being reinterpreted as having arisen from deposition by turbidity currents, and the name turbidites was coined for these rocks. The mechanism of turbidite deposition was pure catastrophism because it implied that there were long intervals of nondeposition or very slow deposition that were periodically interrupted by brief episodes of deposition of an appreciable thickness of sediments.

The concept of gradualism in sedimentation was on the way out, slowly replaced by a view that many sedimentary rocks had been deposited during large-scale depositional events. In retrospect it is difficult to understand why the theory of a slow, gradual accumulation of layered rocks ever dominated. After all, natural events are generally not gradual. A storm is followed by a calm period, which is followed by another storm. The carrying capacity of a river varies widely with the flow, which diminishes to a trickle in some seasons and rises to a raging torrent in others. A single moderate rainstorm can increase a small river's total flow and rate of flow by factors of hundreds or thousands of times, thereby increasing the geologic work that can be done by the river by vastly greater amounts. As a result most of a river's total erosion and transportation of sediments occurs during brief intervals between which there is little or no sediment movement.

The late John Clark, who was at the South Dakota School of Mines at the time, studied the Oligocene continental rocks of the area and noted the presence of numerous coprolites— fossil feces—in the rocks. Clark assumed that a carnivore scat exposed to the air and weathering would not survive for more than about a year. He therefore concluded that the rate of deposition of the coprolite-bearing beds must have been at least great enough to cover a scat in a year. From the size of the coprolites he concluded that the rate of deposition of the sediment was about 2 centimeters (about 0.8 inches) per year. Clark also made the gradualistic assumption that sediment accumulated at an approximately constant rate during the deposition of the entire sequence of rocks. Knowing the rate of sedimentation and the total thickness of the rock unit, he concluded that it took 20,000 years for the unit to be deposited. Extensive geologic and paleontologic evidence now indicates that the rocks he studied were undoubtedly deposited during many brief episodes of rapid deposition alternating with periods of nondeposition or even erosion that encompassed several million years. Numerous brief episodes of deposition account equally well, if not more satisfactorily, for the presence of the numerous coprolites, because rapid deposition of thick layers of rock in a brief span of time would favor preservation of easily decomposed organic matter such as animal droppings. Clark's time estimate is short by a factor of about two hundred. Scatology must be used with caution both in humor and geology.

In the oceans a violent storm can stir up and move huge quantities of sediment or produce extensive erosion of shorelines. During the winter of 1982–1983 storms along the California coast denuded favored bathing beaches of their sand, leaving only cobbles and boulders. It was little comfort to the sunbathers to be told that the sand had been deposited elsewhere, offshore. What the beach bums viewed as a catastrophe was indeed a catastrophic geologic event of erosion at one site and deposition at another, and the rates of accumulation of the sand must have been very large at some locations.

In the 1970s the concept that rocks had a heritage of erosion and deposition during violent storms became more popular, with burrower-expert Adolf Seilacher being one of the prime advocates. According to that theory, a given portion of a sedimentary sequence is the product of deposition during a single storm, as with the sand bodies that were formed off California in 1982–1983. A sequence of rocks of this type is nothing more than a pile of rock units deposited one on top of the other in a single event. Between storms not much happened, but during a tempest these rocks, called tempestites, were formed.

A modern view of geologic history would be that a succession of events took place in an evolving environment so that conditions were never exactly the same as they had been previously. But superimposed on these unidirectional changes inferred from the rocks are also features that suggest recurring or cyclic events.

Scientists and mystics share a common attraction to cyclic events, presumably because of the strong physical and psychological effects of daily cycles of night and day and the seasonal cycles of hot to cold and wet to dry. Longer-term cycles in weather are also well documented. Cycles are also appealing because they offer the hope of predicting the future by examining the past.

Paleontologists have sought evidence for cycles in fossil invertebrates and have proposed that very small-scale layering in some fossils' skeletons record daily cycles in the past. From those cycles they have demonstrated that the rotation of the earth was more rapid in times past. Evidence of annual cycles has been found in both plants and animals, tree rings being a familiar example. Longer cycles that are possibly related to variations in sunspot activity have also been discovered.

Geologists have found cyclic repetition in various rocks. They have recently recognized that fossil soils are common in volcanic rocks and in sediments that were laid down on the continents. Such soils are formed when a unit of fresh rock is overlain by a soil that is in turn overlain by another unit with a soil on top, and so forth. In volcanic sequences a layer of volcanic ash or a lava flow appears to have been added whenever volcanism was at an active stage, with no regular time interval between eruptions. In a similar way, deposition of a sedimentary unit is not thought to be related to any regular external cause but more likely reflects a single storm, a change in a river's course, the melting of a snow-pack in a distant mountain mass, or some other unusual and irregular event. Between depositional events little took place except soil formation through weathering together with greater or lesser erosion of loose material.

There are many marine and mixed marine–continental sedimentary sequences that show repetitions of sedimentary units. As mentioned previously successive underwater mud-slides can build up sequences of so-called turbidites. Many coal-bearing sediments show regular repetitions of rock types, generally reflecting changes in salinity and water depth during their deposition.

Most of the sedimentary sequences in which repetitive sequences have been noted are those made of particles such as silt or sand, the so-called clastic rocks that include shales and sandstones. This is so because the variations in this rock type can be readily observed in the field by almost any geologist. Repetitive sedimentation has also been discovered in recent years in rocks consisting of carbonates—the limestones and dolomites. The nature of these units requires more careful observation of small details than is necessary with shales and sandstones because, superficially, one limestone or dolomite looks just like another. Not that carbonates are really any different from other sedimentary rocks, for they too are made up of particles of calcium carbonate that obey the same physical laws as ordinary sand or silt grains. The particles may be of organic origin, but they are still particles. (I am excluding reef accumulations in which a skeletal framework of hard parts of organisms is preserved.)

Some newly reported work in the early Paleozoic rocks of

New York state is of interest in this regard. E. J. Anderson and Peter Goodwin of Temple University and Barry Cameron of Boston University have studied the carbonates of the area (of Middle Ordovician and Lower Devonian age) and have discovered depositional cycles throughout the rocks that were not recognized by the generations of geologists who have studied those rocks since the early nineteenth century. Like others who have studied cyclic rocks, they found that an individual sedimentary sequence consists of rocks that reflect deposition in successively shallower water as you look upward in the sequence. At the top of each sequence there is a sharp break that represents an interval of nondeposition, including even exposure to the atmosphere and erosion, and then the next sequence begins.

Anderson, Goodwin, and Cameron, as well as geologists who have studied similar sequences in other parts of the world, interpreted the sharp breaks as intervals of rapid transgression of the sea onto the land during which there was no deposition and even small amounts of erosion. They further interpreted their evidence to mean that large-scale transgressions are not a continuous phenomenon but proceed in a series of abrupt episodes of the subsiding of the depositional regions that are in turn the result of large-scale movements in the crust of the earth. In their scheme there is a sudden drop in the altitude of the depositional surface that allows the sea to transgress farther onto the land, causing a deepening of the water at any given point. During this brief deepening episode there is no deposition of rock layers. What follows is a gradual accumulation of sediment over long periods during which the water becomes shallower in a more or less continuous way until the next abrupt deepening.

Such an interpretation is in keeping with the new catastrophism theories of turbidites and tempestites and is likely to be accepted by many in the geologic fraternity. Certainly an interpretation that includes the influence of an outside event such as sudden crustal movements is tempting because it means that the changes in the rocks should be simultaneous within a whole depositional basin. Hence abrupt changes observed in one place could be correlated with those in another even though the exact rock types affected might be very different. Such a time plane would be very convenient in analyzing the New York state rocks because they are made up of a variety of carbonate rocks that were deposited in water of different depths in different parts of a large depositional basin. Geologists would find it useful to be able to correlate a depositional break found in a mudflat deposit with one in a rock sequence deposited in deeper water. If a basinwide effect was triggered by an external event, geologists would have a valuable tool for basinwide correlation in a region where the extent and number of rock exposures are a far cry from the sort of continuity visible in a place like the walls of the Grand Canyon.

Catastrophism is a tempting explanation because it suggests that there is a cause for every effect that can be isolated and identified, not unlike the creationists' crediting the deity with responsibility for everything observed in the geologic record. However, many events may be explained without reference to an external influence. Certainly a layer of volcanic rock is caused by an eruption of ash or lava, but there need be no major external cause for that eruption, which very likely resulted from a slow accumulation of gases or the gradual rise of molten rock from deep in the crust until the overlying rocks yielded and activity finally broke through to the surface. Similarly, a layer of sandstone can be attributed to a single storm or other"unusual" event, but the storm or other event was likely just one in a neverending sequence of events that are largely unrelated either to one another or to a more general cause. There are lots of tempests, so there are lots of tempestites.

Some catastrophic depositional events may well be related to gradual or continuous phenomena. For example, the sudden movement of a mass of wet sediment in the form of a submarine mudflow or turbidity current can be blamed on an earthquake, as it was in 1929 when the transatlantic cables were broken, but that need not be the general rule. A very probable scenario for the turbidite sequences is that sediments were deposited over a long period at the margin of a depositional basin. After enough sediment had been piled up, the weight of the accumulated mass exceeded the strength of the underlying material, and the whole body spontaneously slid away. No earthquake or divine intervention was required. A turbidity current would now and then slide downslope to deposit a layer of turbidite, and the whole process of slowly depositing sediments would recommence, continuing until instability was again reached. The timing of the mudslides is essentially independent of outside events as long as sediment continues to accumulate in the source area for the slides, and in theory at least, periodic slides would take place even if there were perfectly constant external conditions. That's *gradual catastrophism.*

The concept of gradual catastrophism means that the potential for an abrupt geologic event is stored in some way, ready to be released at some future time, either spontaneously or through the intervention of an external event. What takes place physically is the storage of energy in some form. For a simple example consider the earthquake phenomenon. The crust of the earth is subject to slow deformation almost constantly. Some of this deformation is permanent and is reflected by folded rocks and the like. Some of it, though, is stored in the form of elastic, recoverable deformation—elastic energy, if you like. If the elastic strain exceeds the strength of the rock, the rock will fail (break or bend) abruptly, releasing part of the energy as an earthquake. The failure commonly takes place along a preexisting zone of low strength, a fault. The storage of energy as elastic strain is fairly obvious, but energy can be stored in other ways as well.

In a pile of soft sediment that eventually slides away as a turbidity current, the energy is stored as the mass of the sediment, which has the potential to accelerate under the influence of gravity, converting the stored energy into motion. Energy is stored in hills and mountains in the same way. Every particle in a mountain has the potential to move to a lower elevation. When the potential energy of a single particle or a large rock mass exceeds its cohesion to the rest of the mountain it will move downslope spontaneously. Such movements are sometimes triggered by an event like an earthquake, but the triggering only sets off what would have happened eventually anyway.

Mountain masses also act as traps for solar energy stored in the form of snow, and a snowmelt can combine with the rock particles and masses to move very large amounts of material in abrupt events. Even snowmelt is not required, however, because mountains act as barriers that cause heavy rainfalls from moist air. Thus topography can combine with weather patterns to produce massive erosion and deposition events that are certainly catastrophic, as when great floods roar down mountain canyons. Yet these events are self-generating; they are not caused by an external event.

Energy can be stored in organisms as well as rocks, to be reflected later in biological events. A familiar example is the relationship between rabbit and fox populations in which energy is stored in the form of rabbits. That edible energy is released as an increase in the number of foxes, which then use up the energy source so that many foxes die off. This cycle then repeats itself once the rabbit population has recovered. Such energy storage can account for explosive expansions of organisms. For example, wooly mammoths migrated to North America in the late Pliocene and underwent a rapid increase in numbers, after which they died off in the Pleistocene epoch. The mammoths utilized a vast stored energy resource in the form of leaves on trees. The leaves on the upper parts of trees were simply not accessible to any other organisms of quantitative importance so the elephants prospered on the vast food supply, expanding their numbers greatly.

I propose that energy storage also accounts for the cycles observed in the carbonates of Ordovician and lower Devonian age in New York that I mentioned earlier. The geologists who discovered the cycles attributed them to the effect of sudden changes in sea level, with one cycle ending as the sea transgressed the land and a new cycle beginning with rocks characteristic of deep-water depositions. However, it seems unlikely that the basin in which these rocks were deposited was subject to abrupt changes in sea level because it was well within a stable continental mass at the time of deposition of the carbonate sediments. Let us assume instead that the transgression of the sea onto the land was a gradual process, more or less continuous in rate, instead of occurring by steps. Let us also assume that during the deposition of a cyclic carbonate unit the rate of accumulation of sediment was greater than the rate of deepening of the water. In such a situation the water depth would slowly diminish and the character of the sediments would reflect that change. As the water became shallower, the particle size would generally become finer as the transportation capacity of the water became more limited. The finest particles would form mudflats around margins of the basin that with time would extend toward the basin as the basin filled.

Did this process take place basinwide? I think not. If it had, the shores and mudflats would have advanced into the basin from all the directions of shallowing, which is not possible because the sedimentary particles must have had a source. The particles must have been formed and accumulated somewhere and then moved down an energy gradient created by gravity and currents to the site of deposition. The mudflats moved into deeper water away from the source of the particles, not necessarily away from the shore. The depositional hiatus at the end of a cycle in the carbonate rocks represents a time during which there was little or no net movement of particles, or perhaps a net loss of particles. Most likely, the mudflat environment progressed parallel to, as well as away from, the shore. At any given point, sedimentation would cease when gradients became too low to move particles as on the mudflats, or when particles ceased to be supplied for one of two reasons. Either the source became exhausted or the distance became too great relative to the elevation differential—that is, the forces resisting transport became equal to the potential energy difference between the source and the site of deposition.

According to this scheme, carbonate particles would accumulate at some source area until the accumulation was large enough to attain a sufficiently high potential energy level to overcome an internal resistance to erosion, at which time particles could be transported. Particles would then be moved away until a state approaching equilibrium was reached again, thus ending a cycle. Meanwhile, at some other part of the basin, another potential source mass of particles would have been accumulating. At some point it too would start to be transported and another cyclic event would have begun to be added to the sedimentary record, perhaps even reaching out to and overlapping the first one. Each depositional cycle would have begun when the energy difference between a source and a distant depositional site became sufficiently great. At any given place in the basin, the cycle makes its appearance when the first particles from the latest sediment outburst (what we could call a sedimentation front) reach that point. At any given time during basin evolution, there were probably several source areas where particles were being accumulated by some precipitation mechanism, which combined the organic and inorganic secretion of calcium carbonate. At any potential energy difference greater than zero these masses of sediment are not in stable equilibrium with the basin, but they would persist, and even increase in mass, in metastable equilibrium. As the basin continuously sank, the energy difference between the source mass and other parts of the basin would finally reach a critical threshold value and transportation of the mass would begin. In this scheme the tops of the cycles found in the limestones and dolomites transgressed time as they formed at distances increasingly farther from the sources. The cycles are thus self-triggering, the reflection of the storage of energy in accumulations of sediment in one place at a higher energy potential than at some other potential site of deposition. This is not to say that an episode of deposition couldn't be triggered by a violent storm or other external cause, but the potential for movement exists in the accumulation of sediment in a metastable state of excess gravitative energy. An additional factor that could well participate in the sequence of events is vertical movement of the crust in response to movement of the sediment. At the source area the sediment accumulation would have caused some local elastic depression of the crust because of the weight of the sediment. As the source area was depleted, that part of the crust would rebound to help maintain the potential difference between the source and the regions of deposition where, at the same time or soon after, some amount of elastic depression would occur as the weight of sediment increased. At the end of a cycle, near equilibrium would have been reached, and there would be no net transport of sediment as the water continued its slow deepening. This stage of little change would persist until the next sedimentation front arrived from a distant source.

The location of source areas in such a situation is difficult or impossible to unravel. A modern example in which the source is known may serve to clarify some concepts. The delta of the Mississippi River is an accumulation of sediment that has the river as its source. The supply of sediment is not exactly constant but is at least fairly steady. Yet if we were to examine a drill core from the delta, we would find that it revealed cycles of sedimentation, with the bottom of a cyclic unit showing features suggesting deeper water and higher energy of water movement than the top. If the core was from an ancient sandstone, we might propose that the cycle was formed in response to an external cause such as abrupt basin deepening. In the case of the Mississippi delta, we know from historical records that the cycles represent changes in the location of the main channel of the river. Sediment is deposited until the gradient becomes so low that no more sediment can be moved and the river's flow is blocked—plugged by its own sediment load. The supply of sediment-laden water is steady, so the water seeks another, steeper gradient course, and another lobe of the delta is begun. The cycles are nothing more than overlapping lobes of the greater delta, successive waves of deposition from a single source. The energy storage unit is the Mississippi River, a mighty body of water and sediment with a single energy output point and uncounted input points upstream. A catastrophic event in part of the drainage basin of the Mississippi, such as a torrential rain or a gigantic mudslide, does not affect what happens in the delta because the river acts as a giant buffer, ironing out the minor variations and presenting what is essentially a continuum to the delta. Events within the delta result in the cyclic deposition of sedimentary rocks from a constant supply of sedimentary particles.

I could cite many more examples, but the principle is the same. Perhaps it is more useful to look into some generalities about cycles in which energy storage is the cause. Two features of cyclic events are of interest: the frequency of the events and their magnitude. In the case of earthquakes, if a seismologist could predict these two things, he would have done his job very well as far as residents of earthquake-prone regions are concerned. The two factors that affect frequency and magnitude are the capacity for energy storage in the system and the rate at which energy is supplied. Frequency of events is proportional to the rate at which energy is supplied and inversely proportional to the capacity for energy storage. The magnitude of the event is proportional to the capacity for energy storage. We can use a hydraulic analogue to illustrate this concept.

Suppose you had a bucket that was pivoted at a point along the sides that was just below the center of gravity of the bucket. If you put water into the bucket with a hose, it would remain stable as long as the waterline was below the pivot. As the water level climbed, the weight of water above the pivot would finally equal that below and then exceed it, at which point the bucket would no longer be stable and would tip and dump the water. The amount of energy released would be equal to the carrying capacity of the bucket. If the water provided from the hose was reduced, it would take longer to fill the bucket, so the frequency of dumping would be less, but the energy released, per dump, would be the same; it would just happen less often. If the hose valve were opened wide, the bucket would fill more quickly and dump more frequently, but the energy released, per dump, would be a constant. The only way more energy could be released in a single dump would be to use a bigger bucket. At any time, an external cause, such as your tipping the bucket manually, could release whatever amount of energy was stored there at the time. This amount of energy could never be more than the maximum released in a self-perpetuating cycle. Also, the bucket could be held to prevent its slumping and the water would continuously overflow at a slow rate, but the potential for rapid release as soon as your hand was removed would always be there.

In geology it seems that cycles are found everywhere you seek them, and it seems clear why. Energy storage is something that takes place in a variety of ways and in numerous forms. The various storage systems are provided with energy in the form of heat from the interior of the earth or radiant energy from the sun. The energy may not appear in the form of heat or radiation, but that's where it comes from. The earth's heat may appear as motion of crustal masses and solar energy may appear as vaporized water, but the two fundamental energy sources remain the same. Both sources provide energy constantly, so we should expect that this steady flow of energy would have been stored in a multitude of ways both in the past and in the present. Where there is energy storage, there are cyclic releases of that energy that are reflected in geological events.

Catastrophies are just cyclic events that result from large energy storage capacity. Gradualism as a principle applies only to the way energy is provided to the system—earth—at a more or less steady rate. The result of this gradualism is necessarily always cyclic and frequently catastrophic. Catastrophes are forever.

Appendix A

GEOLOGIC TIME SCALE*

interval name	approximate duration, millions of years before present	% of total time	events, life forms
PHANEROZOIC TIME		12	
CENOZOIC ERA		1.4	
Quaternary Period			
Holocene Epoch	0.01–0.0	0.0002	modern man
Pleistocene Epoch	2.0–0.01	0.04	early man
Tertiary Period			
Pliocene Epoch	5.1–2.0	0.06	hominids
Miocene Epoch	24.6–5.1	0.41	Himalayan Mountains formed
Oligocene Epoch	38–24.6	0.28	
Eocene Epoch	54.9–38	0.35	early horses
Paleocene Epoch	65–54.9	0.21	early primates
MESOZOIC ERA		3.8	
Cretaceous Period	144–65	1.6	last of dinosaurs; Alps, Rockies formed

*Numerical values from Harland and others, 1982.

GEOLOGIC TIME SCALE (*Continued*)

interval name	approximate duration, millions of years before present	% of total time	events, life forms
Jurassic Period	213–144	1.4	dinosaurs, early birds, and mammals
Triassic Period	248–213	0.73	Atlantic Ocean formed
PALEOZOIC ERA		7.1	
Permian Period	286–248	0.79	reptiles
Carboniferous Period*	360–286	1.5	early reptiles, trees
Devonian Period	408–360	1.0	early trees
Silurian Period	438–408	0.62	early land plants
Ordovician Period	505–438	1.4	first fish
Cambrian Period	590–505	1.8	first shelled organisms
PRECAMBRIAN TIME		88	
PROTEROZOIC ERA	2,450–590	39	first complex organisms, oxygen in atmosphere
ARCHEAN ERA AND PRISCOAN ERA	4,800–2,450	49	bacteria and algae in last half of era

*Within the United States the Carboniferous Period is called the Carboniferous System and is subdivided into an older Mississippian Period and a younger Pennsylvanian Period.

Appendix B

THE CHEMICAL ELEMENTS

(Omitting noble gases and short-lived, radioactive elements)

element	sym-bol	atomic number	average mass number	approximate abundance in crust (ppm)	
				Mason*	Vino gradov*
hydrogen	H	1	1.008	1,400	—
lithium	Li	3	6.94	20	32
beryllium	Be	4	9.01	3	—
boron	B	5	10.81	10	12
carbon	C	6	12.01	200	230
oxygen	O	8	15.99	466,000	470,000
fluorine	F	9	18.99	625	660
sodium	Na	11	22.99	28,300	25,000
magnesium	Mg	12	24.31	20,900	18,700
aluminum	Al	13	26.98	81,300	80,500
silicon	Si	14	28.09	277,200	295,000
phosphorus	P	15	30.97	1,050	930
sulfur	S	16	32.06	260	470
chlorine	Cl	17	34.45	130	170

potassium	K	19	39.10	25,900	25,000
calcium	Ca	20	40.08	36,300	29,600
scandium	Sc	21	44.96	22	10
titanium	Ti	22	47.90	4,400	4,500
vanadium	V	23	50.94	135	90
chromium	Cr	24	51.99	100	83
manganese	Mn	25	54.94	950	1,000
iron	Fe	26	55.85	50,000	46,500
cobalt	Co	27	58.93	25	18
nickel	Ni	28	58.71	75	58
copper	Cu	29	63.54	55	47
zinc	Zn	30	65.37	70	83
gallium	Ga	31	69.72	15	19
germanium	Ge	32	72.59	2	1
arsenic	As	33	74.92	2	2
selenium	Se	34	78.96	0.05	0.05
bromine	Br	35	79.91	2	2
rubidium	Rb	37	85.47	90	150
strontium	Sr	38	87.62	375	340
yttrium	Y	39	88.90	33	29
zirconium	Zr	40	91.22	165	170
niobium	Nb	41	92.91	20	20
molybdenum	Mo	42	95.94	1	1

THE CHEMICAL ELEMENTS (*Continued*)
(Omitting noble gases and short-lived, radioactive elements)

element	sym-bol	atomic number	average mass number	approximate abundance in crust (ppm)	
				*Mason**	*Vino gradov**
ruthenium	Ru	44	101.07	0.01	—
rhodium	Rh	45	102.91	0.005	—
palladium	Pd	46	106.4n	0.01	0.01
silver	Ag	47	107.87	0.1	0.1
cadmium	Cd	48	112.40	0.2	0.1
indium	In	49	114.82	0.1	0.2
tin	Sn	50	118.69	2	2
antimony	Sb	51	121.75	0.2	0.5
tellurium	Te	52	127.60	0.01	0.001
iodine	I	53	126.90	0.5	0.4
cesium	Cs	55	132.90	3	—
barium	Ba	56	137.34	425	650
lanthanum	La	57	138.91	30	29

rare earths—Ce, Pr, Nd, Sm, Eu, Gd, Tb, Dy, Ho,
Er, Tm, Yb, Lu—nos. 58–71, at. wts. 140.12–174.97.

			<60	<70
hafnium	Hf	72	178.49	3	1
tantalum	Ta	73	180.95	2	2
tungsten	W	74	183.85	1	1
rhenium	Re	75	186.2n	0.001	0.001
osmium	Os	76	190.2n	0.005	—
iridium	Ir	77	192.2n	0.001	—
platinum	Pt	78	195.09	0.01	—
gold	Au	79	196.97	0.004	0.004
mercury	Hg	80	200.59	0.08	0.08
thallium	Tl	81	204.37	0.5	1
lead	Pb	82	207.19	13	16
bismuth	Bi	83	208.98	0.2	0.1
thorium	Th	90	232.04	7	13
uranium	U	92	238.04	2	2

n indicates no more significant figures

*Abundance values from Brian Mason, *Principles of Geochemistry* (New York: Wiley, 1966); and A. P. Vinogradov, "Average Contents of Chemical Elements in the Principal Types of Igneous Rocks of the Earth's Crust," in *Geochemistry* (translation of *Geochemiya* 1962, Issue #7), p. 641–664.

Further Reading

1 ADAMS, FRANK DAWSON. *The Birth and Development of the Geological Sciences*. 1938. Reprint. New York: Dover, 1954.

2 ANDERSON, E. J., CAMERON, B., and P. W. GOODWIN. *Helderberg Punctuated Aggradational Cycles*. 1980. Eastern Section, Society of Economic Paleontologists and Mineralogists, Guidebook 1980.

3 ANDERSON, E. J., PETER W. GOODWIN, and THEODORE H. SOBIESKI, 1984, "Episodic Accumulation and the Origin of Formation Boundaries in the Helderberg Group of New York State." *Geology*, v. 12, p. 120–123.

4 BAGNOLD, R. A. *The Physics of Blown Sand and Desert Dunes*. 1941. Reprint. London: Chapman and Hall, 1973.

5 BARTH, TOM. F. W. "Abundance of the Elements, Areal Averages and Geochemical Cycles." *Geochem. et Cosmochem. Acta*. 23 (1961): 1–8.

6 ———. "Ideas on the Interrelations of Igneous and Sedimentary Rocks." *Bull. Comm. Geol. Finlande* 196 (1961): 321–326.

7 BEERBOWER, JAMES R. "Cyclothems and Cyclic Depositional Mechanisms in Alluvial Plain Sedimentation." *Geol. Survey Kansas Bull.* 169 (1964): 31–42.

8 BEHRENSMEYER, ANNA K. "Time Resolution in Fluvial Vertebrate Assemblages." *Paleobiology* 8 (1982): 211–227.

9 BEHRENSMEYER, ANNA K., and ANDREW P. HILL, eds. *Fossils in the Making: Vertebrate Taphonomy and Paleoecology.* Chicago: University of Chicago Press, 1980.

10 BERRY, WILLIAM B. N. *Growth of a Prehistoric Time Scale.* San Francisco: W. H. Freeman, 1968.

11 BIRCH, FRANCIS. "Speculations on the Earth's Thermal History." *Geol. Soc. America Bull.* 76 (1965): 133–154.

12 BOHOR, B. F., E. E. FOORD, P. J. MODRESKI, and D. M. TRIPLEHORN. "Mineralogic Evidence for an Impact Event at the Cretaceous-Tertiary Boundary." *Science* 224 (1984): 867–869.

13 BOTT, MARTIN H. P. *The Interior of the Earth: Its Structure, Constitution and Evolution.* 2d. ed. New York: Elsevier, 1982.

14 BOWEN, H. J. M. *Trace Elements in Biochemistry.* New York: Academic Press, 1966.

15 BROWN, A. B. "Bone Strontium as a Dietary Indicator in Human Skeletal Populations." *Contrib. to Geology* 13 (1974): 47–48.

16 CLARKE, F. E. *The Data of Geochemistry.* U.S. Geol. Survey, Bull. 770. Washington, D.C.: Government Printing Office, 1924.

17 CLARKE, F. E., and H. W. WASHINGTON. *The Composition of the Earth's Crust.* U.S. Geol. Survey, Prof. Paper 127. Washington, D.C.: Government Printing Office, 1924.

18 CLOUD, P. "The Primitive Earth." In *Understanding the Earth,* edited by Gass et al. Artemis Press, 1971.

19 COLE, CHARLES J. "Chromosome Inheritance in Parthenogenetic Lizards and Evolution of Allopolyploidy in Reptiles." *Journal of Heredity* 70 (1981): 95–102.

20 ———. "Unisexual Lizards." *Scientific American* 250, no. 1 (1984) 94–100.

21 CURTIS, G. H., J. F. EVERNDEN and J. LIPSON. *Age Determi-*

nations of Some Granitic Rocks in California by the Potassium-Argon Method. 1958. Calif. Div. Mines Spec. Rept. 54.

22 DARWIN, CHARLES. *On the Origin of Species.* London: J. Murray, 1869.

23 DOUGALL, H. W., and D. L. W. SHELDRICK. "The Chemical Composition of a Day's Diet of an Elephant." *East African Wildlife Jour.* 2 (1964): 51–58.

24 DUFF P. McL. D., A. HALLAM, and E. K. WALTON. *Cyclic Sedimentation.* New York: Elsevier, 1967.

25 ECHLIN, P. "The Origins of Plants." In *Phytochemical Phylogeny.* New York: Academic Press, 1970.

26 EDMOND, JOHN M., and KAREN VON DAMM. "Hot Springs on the Ocean Floor." *Scientific American* 248, no. 4 (1983): 78–93.

27 EICHER, DON L. *Geologic Time.* 2d ed. Englewood Cliffs, New Jersey: Prentice-Hall, 1976.

28 FERM, JOHN C. "Allegheny Deltaic Deposits." In *Deltaic Sedimentation: Modern and Ancient,* edited by J. P. Morgan. 1970. Special Pub. 15, Soc. of Economic Paleontologists and Mineralogists, pp. 246–255.

29 GOLDSCHMIDT, V. M. *Geochemistry.* New York: Oxford University Press, 1958.

30 GOULD, STEPHEN JAY. "Is Uniformitarianism Necessary?" *Amer. Jour. Sci.* 263 (1965): 223–228.

31 ———. *The Panda's Thumb.* New York: W. W. Norton, 1980.

32 ———. *Hen's Teeth and Horse's Toes.* New York: W. W. Norton, 1983.

33 ———. "Balzan Prize to Ernst Mayer." *Science* 223 (1984) 255–257.

34 GOULD, STEPHEN JAY, and NILES ELDRIDGE. "Punctuated Equilibria: The Tempo and Mode of Evolution Reconsidered." *Paleobiology* 3 (1977): 115–151.

35 HALLAM, A. *Great Geological Controversies.* Oxford: Oxford University Press, 1983.

36 HARLAND, W. B., A. V. COX, P. G. LLEWELLYN, C. A. G. PICKTON, A. G. SMITH, and R. WALTERS. *A Geologic Time Scale.* Cambridge: Cambridge University Press, 1982.

37 HECKEL, PHILIP H., LAWRENCE L. BRADY, W. JAMES EBANKS, JR., and ROGER K. PABIAN. *Field Guide to Pennsylvanian Cyclic Deposits in Kansas and Nebraska.* 1979. Kansas Geol. Survey, Guidebook 4. (Pages of interest are pp. 19–21, presumably by Heckel.)

38 HOLLAND, HEINRICH D. "The Geologic History of Sea Water—An Attempt to Solve the Problem." *Geochem. et Cosmochem. Acta* 36 (1972): 637–651.

39 HOFFMAN, H. J. *Attributes of Stromatolites.* 1969. Geol. Survey Canada, Paper 69–39.

40 HUNT, ROBERT M., JR., XUE XIANG-XU, and JOSHUA KAUFMAN. "Miocene Burrows of Extinct Bear Dogs: Indication of Early Denning Behavior of Large Mammalian Carnivores." *Science* 221 (1983): 364–366.

41 HUTTON, JAMES. *Theory of the Earth; or An Investigation of the Laws Observable in the Composition, Dissolution, and Restoration of Land upon the Globe.* Transactions of the Royal Society of Edinburgh, Vol. 1, Pt. II, pp. 209–304. Edinburgh, 1788. (Paper read in 1785.)

42 JANIS, CHRISTINE, "The Evolutionary Strategy of the Equidae and the Origins of Rumen and Cecal Digestion." *Evolution* 30 (1976): 757–774.

43 JOLY, JOHN. "The Age of the Earth." *Scientific Monthly* 16 (1923): 205–216.

KELVIN, LORD (See William Thomson.)

44 KERR, PAUL F. "Quick Clays." *Scientific American* 209, no. 5 (1963): 132–142.

45 LI, YUAN-HUI. "Geochemical Mass Balance among the Lithosphere, Hydrosphere, and Atmosphere." *Am. Jour. Sci.* 272 (1972): 119–137.

46 LIVINGSTONE, D. A. "The Sodium Cycle and the Age of the Ocean." *Geochem. et Cosmochem. Acta* 27 (1963): 1055–1069.

47 LYELL, CHARLES. 1830–1833. *Principles of Geology.* 3 vols. London: J. Murray. (1969 reprint by Johnson Reprints, New York.)

48 MASON, BRIAN. *Principles of Geochemistry.* 3d ed. New York: Wiley, 1966.

49 MATTHEW, W. D. "Fossil Bones in the Rock: The Fossil Quarry near Agate, Sioux County, Nebraska." *Natural History* 23 (1923): 359–369.

50 MITCHELL, A. H. G., and M. S. GARSON. *Mineral Deposits and Global Tectonic Settings.* New York: Academic Press, 1981.

51 McCONNELL, D. "Dating of Fossil Bones by the Fluorine Method." *Science* 136 (1962): 241–244.

52 MORGAN, JAMES P. "Depositional Processes and Products in the Deltaic Environment." In *Deltaic Sedimentation: Modern and Ancient,* edited by J. P. Morgan. 1970. Special Pub. 15, Soc. of Economic Paleontologists and Mineralogists, pp. 31–47.

53 NUTTLI, OTTO W. "The Mississippi Valley Earthquakes of 1811 and 1812: Intensities, Ground Motion, and Magnitudes." *Seismological Soc. America Bull.* 63 (1973): 227–248.

54 OAKLEY, K. P. "The Fluorine Dating Method." In *Yearbook of Physical Anthropology (1949).* 1950, pp. 44–52.

55 OAKLEY, K. P., and C. R. HOSKINS. "New Evidence on the Antiquity of Piltdown Man." *Nature* 165 (1950): 379–382.

56 O'CONNELL, R. J., and B. H. HAGER. "On the Thermal State of the Earth." In *Physics of the Earth's Interior,* edited by A. M. Dziewonski and E. Boschi. Amsterdam: North Holland, 1980, pp. 270–317.

57 PARKER, RONALD B., "Electron Microprobe Analysis of Fossil Bones and Teeth," 1967. *Geol. Soc. America Spec. Paper* 101, p. 415.

58 PARKER, RONALD B., J. W. MURPHY, and HEINRICH TOOTS. "Fluorine in Fossil Bone and Tooth: Distribution among Skeletal Tissues." *Archaeometry* 16 (1974): 98–102.

59 PARKER, RONALD B., and HEINRICH TOOTS. "Minor Elements in Fossil Bones." *Geol. Soc. America Bull.* 81 (1970): 925–932.

60 ———. "A Final Kick at the Fluorine Dating Method." *Arizona Acad. Sci. Journal Proc.* 11 (1976): 9–10.

61 ———. "Sodium in Fossil Vertebrates as a Paleobiological Tool." *Ariz. Acad. Sci. Journal Proc.* 11 (1976): 87.

62 ———. "Trace Elements in Bones as Paleobiological Indi-

cators." In *Fossils in the Making,* edited by A. K. Behrens-meyer and A. P. Hill. Chicago: University of Chicago Press, 1980, pp. 197–207.

63 PARKER, RONALD B., HEINRICH TOOTS, and J. W. MURPHY. "Leaching of Sodium from Skeletal Parts during Fossilization." *Geochem. et Cosmochem. Acta* 38 (1974): 1317–1321.

64 PENICK, JAMES L., JR. *The New New Madrid Earthquakes.* Columbia, Missouri: University of Missouri Press, 1981.

65 PLAYFAIR, J., *Illustrations of the Huttonian Theory.* Edinburgh: 1802. (1956 reprint by University of Illinois Press, Urbana.)

66 POLDERVAART, ARIE. "Chemistry of the Earth's Crust." In *Crust of the Earth,* edited by Arie Poldervaart. 1955. Geol. Soc. America Spec. Paper 62.

67 READ, H. H., and JANET WATSON. *Introduction to Geology.* Vol. 2, Part 1, *Earth History: Early Stages in Earth History.* New York: Halsted Press, 1975.

68 RONOV, A. B., and A. A. YAROSHEVSKY. "Chemical Composition of the Earth's Crust." In *The Earth's Crust and Upper Mantle,* edited by P. J. Hart. 1969. Amer. Geophysical Union Monograph 13, pp. 37–57.

69 ROSENTHAL, H. L. "Uptake, Turnover, and Transport of Bone-Seeking Elements in Fishes." *N.Y. Acad. Sci. Annals* 109 (1963): 278–293.

70 RUBEY, WILLIAM W. "Geologic History of Sea Water: An Attempt to State the Problem." *Geol. Soc. America Bull.* 62 (1951): 1111–1148.

71 SCHOENINGER, MARGARET J., 1979, Dietary Reconstruction at Chalcatzinga, A Formative Period Site in Morelos, Mexico: Museum of Anthropology, Univ. of Michigan, Technical Report 9.

72 SCHOENINGER, MARGARET, MICHAEL J. DENIRO, and HENRIK TAUBER. "Stable Nitrogen Isotope Ratios of Bone Collagen Reflect Marine and Terrestrial Components of Prehistoric Human Diet." *Science* 220 (1983): 1381–1383.

73 *Scientific American* 249, no. 3 (1983). (This issue presents the concept of equilibrium cycles in earth processes.)

74 SCOTT, RONALD F. *Principles of Soil Mechanics.* Reading, Mass.: Addison-Wesley, 1963.

75 SEDERHOLM, J. J. "The Average Composition of the Earth's Crust in Finland." *Bull. Commission Geol. Finlande* 12, no. 70 (1925).

76 SHELDON, R. W., and S. R. KERR. "The Population Density of Monsters in Loch Ness." *Limnology and Oceanography* 17 (1972): 796–798.

77 SHELFORD, V. E. *The Ecology of North America.* Urbana, Ill.: Illinois University Press, 1964.

78 SKINNER, BRIAN. "A Second Iron Age Ahead." *American Scientist* 64 (1976): 258–269.

79 TERZAGHI, KARL, and RALPH PECK. *Soil Mechanics in Engineering Practice.* 2d ed. New York: Wiley, 1967.

80 THOMSON, WILLIAM. "On the Age of the Sun's Heat." *Macmillans Magazine* 5 (1862): 288–375.

81 ———. "On the Secular Cooling of the Earth." *Royal Soc. Edinburgh Proc.* 4 (1862): 610–611.

82 ———. "On the Elevation of the Earth's Surface Temperature Produced by Underground Heat." *Royal Soc. Edinburgh Proc.* 5 (1864): 200–201.

83 ———. "On the Secular Cooling of the Earth." *Royal Soc. Edinburgh Trans.* 23 (1864): 157–170.

84 ———. "The 'Doctrine of Uniformity' in Geology Briefly Refuted." *Royal Soc. Edinburgh Proc.* 5 (1865): 512–513.

85 THURBER, D. L., J. L. KULP, E. HODGES, P. W. GAST, and J. M. WAMPLER. "Common strontium content of the human skeleton." *Science* 128 (1958): 256–257.

86 TOOTS, HEINRICH. "The Chemistry of Fossil Bones from Wyoming and Adjacent States." *Contrib. to Geology* 2 (1963): 69–80.

87 TOOTS, HEINRICH, and RONALD B. PARKER, "Thallium in Salt Substitutes: A Possible Health Hazard." *Environmental Research* 14 (1977): 327–328.

88 TOOTS, HEINRICH, and M. R. VOORHIES. "Strontium in Fossil Bones and the Reconstruction of Food Chains." *Science* 149 (1965): 854–855.

89 VERHOOGEN, JOHN, FRANCIS J. TURNER, LIONEL E. WEISS,

CLYDE WAHRHAFTIG, and WILLIAM S. FYFE, *The Earth.* New York: Holt, Rinehart and Winston, 1970.

90 VINOGRADOV, A. P. "Average Contents of Chemical Elements in the Principal Types of Igneous Rocks in the Earth's Crust. In *Geochemistry* (translation of *Geochemiya*), 1962, Issue #7, pp. 641–664.

91 VOORHIES, M. R. "Sampling Difficulties in Reconstructing Late Tertiary Mammalian Communities." *Proc. North American Paleontological Convention* (1969): pp. 454–468.

92 ―――. "Taphonomy and Population Dynamics of an Early Pliocene Vertebrate Fauna, Knox County, Nebraska." 1969. *Contributions to Geology,* Special Paper 1.

93 ―――. "Ancient Ashfall Creates a Pompeii of Prehistoric Animals." *National Geographic* 159, no. 1 (1981): 67–75.

94 VOORHIES, M. R., and JOSEPH R. THOMASSON. "Fossil Grass Anthoecia within Miocene Rhinoceros Skeletons: Diet in an Extinct Species." *Science* 206 (1979): 331–333.

95 WEDEPOHL, K. H. "Die Zusammensetzung der Erdkruste." *Fortschritte der Mineralogie* 46 (1969): 145–174.

96 WEIR, J. S. "Spatial Distribution of Elephants in an African National Park in Relation to Environmental Sodium." *Oikos* 23 (1972): 1–13.

97 WILSON, J. T. "Static or Mobile Earth: The Current Scientific Revolution." *Proc. American Philos. Soc.* 112 (1969): 309–320.

98 WOLKOMIR, RICHARD. "Dirty Work." *Technology Illustrated* 3, no. 8 (1983): 54–60.

Index